Reliability Analysis of Composite Power Systems Using FACTS Controllers

Reliability Analysis of Composite Power Systems Using FACTS Controllers

Dr. Suresh Kumar Tummala

Professor & Head, EEE Dept., GMR Institute of Technology

2015

Copyright © 2015 by Dr. Suresh Kumar Tummala

All rights reserved. This book or any portion thereof may not be reproduced or used in any manner whatsoever without the express written permission of the publisher except for the use of brief quotations in a book review or scholarly journal.

First Printing: 2015

ISBN 978-1-329-65408-2

Dedication

To my Guru, Family Members, Collegeues & Friends

Contents

Acknowledgements ix
Preface 2
Chapter 1: Introduction 3
Chapter 2: 11
Chapter 3: 29
Chapter 4: 41
Chapter 5: 63
Chapter 6: 74
Chapter 7: 92
Chapter 8: 114
Appendix 1 116
References 119

Acknowledgements

I would like to thank my teachers, my collegues, and my friends without whose help this book would never have been completed.

I would also thank the Management, Principal and all Faculty & Staff of GMR Institute of Technology, Rajam for their timely support in bringing this book.

Last but not the least, I would thank my family members for their constant encouragement and for their timely support in bringing this book.

Preface

Reliability analysis of Composite Power System mainly deals with the interaction of transmission and generation system. The evaluation of composite system reliability is extremely complex as it is necessary to include detailed modeling of both generation and transmission facilities and their auxiliary elements. It should be noticed that the reliability assessment of composite power systems requires a complete understanding of the engineering implications in the system, the way it operates, the criteria for success, and the possible modes of failure

In Chapter 1&2, Reliability analysis of transmission line with TCSC and Series Compensator has been discussed with all illustrations and algorithms. State space analysis and Series – Parallel representation of the combinations are also presented.

In Chapter 3, Reliability analysis of transmission line with UPFC and Series Compensator has been discussed with all illustrations and algorithms. State space analysis and Series – Parallel representation of the combinations are also presented. A comparison has been carried for the FACTS components of TCSC and UPFC for the given system.

In Chapter 4, Reliability analysis of 6 bus RBTS with different UPFC modules has been discussed with all illustrations and algorithms. Comparison between these modules is presented. System Indices (BPSD, BPII, BPECI), probability of failure and EENS are also presented in both numerical and graphical form.

In Chapter 5, Reliability analysis of 6 bus RBTS with different TCSC modules has been discussed with all illustrations and algorithms. Comparison between these modules is presented. System Indices (BPSD, BPII, BPECI), probability of failure and EENS are also presented in both numerical and graphical form.

In Chapter 6, a comparison of reliability analysis, system indices, probability of failure and EENS has been carried out between UPFC and TCSC of a 6 Bus RBTS in order to find out which component has the best performance values in all aspects.

In Chapter 7, Reliability analysis of 6 bus RBTS with TCSC and UPFC has been discussed with all illustrations and algorithms. State space analysis and Series – Parallel representation of the combinations are presented. Comparison between these modules is presented. System Indices (BPSD, BPII, BPECI), probability of failure and EENS are also presented in both numerical and graphical method. Reliability analysis of IEEE 24 bus RTS was also discussed in this chapter. System Indices, Probability of failure and EENS are calculated and presented in both numerical and graphical method.

In Chapter 8, Conclusion has been made on the entire work and scope for future work is also presented in this chapter.

Chapter 1

1.1 Introduction:

Modern day society would like electric energy to be continuously available on demand. It is, however, neither technically nor economically feasible to plan, construct and operate a power system that has zero likelihood of failure. A basic objective, therefore, is to satisfy the system load requirements as economically as possible and with a reasonable assurance of continuity and quality. This creates the difficult problem of balancing the need for continuity of power supply and the cost involved. However, no matter how much money, time and effort are invested, and how advance are the techniques utilized, it is impossible to eliminate the possibility of equipment failures and the need to remove equipment from service to perform preventive maintenance.

Composite Power System Reliability [73] mainly deals with the interaction of transmission and generation system. The evaluation of composite system reliability is extremely complex as it is necessary to include detailed modeling of both generation and transmission facilities and their auxiliary elements. It should be noticed that the reliability assessment [62] of composite power systems requires a complete understanding of the engineering implications in the system, the way it operates, the criteria for success, and the possible modes of failure.

One of the most basic elements in power system planning is the determination of how much generation capacity is required to give a reasonable assurance of satisfying the load requirements. In addition to providing the means to move the generated energy to the terminal stations, the bulk transmission facilities must be capable to maintain adequate voltage levels and loadings within the thermal limits of individual circuits and also maintain system stability limits. The total problem of assessing the adequacy of the generation and transmission system with regard to provide a dependable and suitable supply at the terminal stations can be designated as composite power system reliability evaluation. The basic function of a composite generation and transmission system is to generate the required electricity and deliver it to the major load points [61].

In this book, two systems are considered for analyzing Reliability indices. One is Roy Billiton Test System which is a 6 Bus system and second is IEEE Reliability Test System which is a 24 Bus System.

Roy Billinton Test System (RBTS) is a simple system which includes sufficient number of Generating Units & Load Points, Double Circuit Transmission Lines, for determining Reliability Indices Efficiently. RBTS have been used extensively by researchers, as a bench mark system, for reliability assessment and other developments in the field of probabilistic applications in power systems. In the past few decades the electric power system has seen tremendous growth both in the terms of size and technology. Such developments mainly include wind energy, increased use of HVDC transmission, integration of Plug in Hybrid Electric Vehicles (PHEV) and Electric Vehicles (EV) and the state-of-the-art communication systems applied to power systems.

The reliability assessment of composite generation and transmission systems has been received considerable attention during the past few years. The main reason is that, as utilities are finding it increasingly necessary to quantitatively evaluate Individual Busbar (For a given Power System, the bus bar which connects the generators (or) which interlinks transmission lines (or) which supplies load to the distribution system are referred as Individual Bus bar) and Overall System Reliability Indices (Intended Reliability Indices of Individual Bus bars are in concert to calculate single point Reliability Indices which is referred as Overall System Reliability Indices). Although these systems cause customer interruptions or loss of service [50] quality very infrequently, they can have a major and widespread effect on society and the environment when problems arise [49]. Reliability evaluation of a composite generation and transmission system is concerned with the problem of determining the adequacy of the combined generation and transmission system with respect to provide a dependable and suitable supply at the bulk load points. Composite system adequacy assessment [44] is very complex since it involves not only system analyses but also many practical considerations.

1.2 Literature Survey:

The bibliography on the application of probability methods and selected papers on the subject of power system reliability evaluation were presented [9, 23, 49, 54]. In continuation with the above bibliography on the subject of composite system reliability [13] and closely related topics, papers in such areas as: probabilistic load flow [2,5,34] probabilistic production costing, probabilistic transient stability evaluation etc. have not been included except where they specifically address power system reliability evaluation.

Updates in an electrical reliability modeling process and an additional new method for evaluating power system design options were described by Propst and Doan [48]. Additional modelings have been developed for evaluating typical electrical systems found in petroleum and chemical facilities. Su et. al [56] presented a method to optimize and dispatch the spinning reserve for total value of a composite power system. They employed the DC power flow and determined generation shift factor so that power generation can be re-dispatched economically and lines transmission congestion can be alleviated effectively.

In recent years, a large demand has been placed on the transmission network, and demands will continue to increase due to an increasing number of nonutility generators and intensified competition among them. Increasing transmission capacity requirements can be achieved by either constructing new transmission facilities. Generally, it is not economically viable to attempt to develop transmission system just by installing new transmission lines because of a variety of environmental, land use, and regulatory requirements. Dalton et. al [26] presented a new value based reliability planning process proposed for planning of a transmission system. Data in system reliability worth determination was presented by Pandey and Billinton [36].

Chowdhury and Koval [41, 45] presented an approach to establish customer responsive transmission system reliability performance standards that provide a measure against which the different planning and design procedures can be evaluated to assist in achieving the primary objective of providing reliable electricity with competitive rates to utility customers. The authors have identified some of the limitations of traditional utility transmission system planning criteria and the possible implications for industrial and commercial customers. Different features of such a reliability model used to compute the reliability performance of a practical transmission system is also one of the research studies extended by the authors [68].

Goel and Feng [39] described a probabilistic method designated as system well-being analysis for evaluating the effect of peak load, load factor, load curtailment philosophy and percentage load curtailed which incorporates the conventional risk index as well as the accepted deterministic criteria identified as being in the healthy and marginal states. Supervisory Control and Data Acquisition (SCADA) system reliability in terms of its expected, aggregate contribution to load curtailment on the power system is also presented. Expressing this aggregate in system minutes and applying an appropriate damage cost function then provides an annual cost measure of SCADA system reliability worth was presented by Bruce [32]. Li [53] presented a method to incorporate aging failures in power system reliability evaluation. It includes development of a calculation approach with two possible probability distribution [11] models for unavailability of aging failures and implementation in reliability evaluation [10]. The defined unavailability of aging failures has a consistent form which is the same as that for repairable failures. Jensen et. al [38] examined a probabilistic network planning program, which include probabilistic models for the load and for the transmission system as well.

Tran et. al [59] presented the key problems of available features and operation modes of the Transmission Reliability Evaluation for Large Scale System (TRELSS) version .2, for assessing reliability indices. These studies were extended to; suggest that the some important input parameters of the TRELSS can be determined optimally from this sensitivity analysis for higher reliability level operation of a system.

Yang et. al [69] proposed a systematic methodology that is based on a breaker oriented system network model including detailed substation configurations and protection system scheme to evaluate bulk power system reliability considering the impact of protection system failures. Wang et. al [57] discussed the application and accuracy of different analysis techniques supporting the determination of industrial and commercial power system reliability and availability.

Mallard and Thomas [1] presented a method applying probability techniques for analyzing the reliability of transmission system. The method considers generation and transmission equipment performance, weather conditions, load cycles, generation dispatch, interconnections and the effect of scheduled outages. The methods available for transmission system reliability evaluation [6] and the conditional probability approach described earlier by the authors were extended by Bhavaraju and Billinton [3] to include models of load and component states. The method minimizes the investment budget for constructing new transmission lines subject to probabilistic reliability criteria, which consider the uncertainties of transmission system elements. Choi et. al [64] presented a method for evaluating nodal probabilistic congestion and reliability indices of transmission systems. Quantitative evaluation of transmission system reliability is very important because successful operation of an electric power system in electricity market depends on transmission system reliability management [66].

A methodology [55] for calculating the reliability indices of a bulk power system using the state enumeration approach which utilizes topological analysis to determine the contribution of each system state to the frequency and duration indices at both the system and the bus level are presented by Satish and Billinton [27,28]. They have illustrated the effect on the expected annual system outage cost. The probabilistic wheeling capability of an interconnected power transmission system was evaluated by Mijuskovis [24]. The calculation of reliability indices of wheeling enables more adequate estimation of the wheeling rate for power transactions between utilities. Special difficulties involved in bulk power system reliability evaluation computational techniques and date requirements are also reviewed [24]. Ringlee et. al [22] presented the issue where and how applications of bulk power indices will be made in the future and what steps may have to be taken to arrive at that point. It offers an overview of trends in performance monitors and short and long term planning procedures to aid the preparation of a forecast of future applications of reliability indices.

A set of investigations about the bulk reliability performance evaluation of the IEEE Reliability Test System were presented by Pinheiro et. al [31]. Several bulk reliability system indices representing a Hierarchical Level II assessment of the new system are provided in the survey. The reliability criteria presently used by Canadian utilities [12] in regard to planning, operating and generating capacities in which the case of adequacy assessment, a comparison of the different utility criteria of IEEE Reliability Test System is reviewed in [37]. The second stage of an activity sponsored by the IEEE application of probability methods subcommittee centered on identifying techniques and approaches for monitoring, measuring, predicting and applying reliability indices [74] in the planning and operation of power systems was discussed by Allan et. al [14]. Billinton and Li [58] described that there is not much published material available dealing with the effects of using multistate generating unit representation in composite system adequacy assessment. They illustrated these effects by the application to the IEEE Reliability Test System. Load point and system indices for the test system are presented to illustrate the impact of incorporating multi state representations in composite system adequacy assessment [7,8]. Advantages of the proposed method are illustrated by an application to the IEEE Reliability Test System (RTS).

The reliability assessment of composite power systems with a very general and integral computer program based on the analytical approach in which the program allows complex reliability analysis of composite systems considering different impacts, effects and policies which can influence the system behavior and its operating conditions were described by Skuletic and Balota [62, 78]. Reliability is only one aspect of the quality of electricity supply. The other aspects, which will not be analyzed here, but which are also very important for a normal exploitation of the power system, are voltage, frequency and harmonics. The approach to composite power system reliability [30] evaluation based on contingency enumeration and a simulation process to access the system state using an AC load flow model in which the analysis starts with a nonlinear optimization procedure to determine system loadability was described by Sharef and Berg [21]. Audomvongseree and Arporn [42] presented an evaluation method based on AC equivalent in which the method proposed categories a system into three main sections which will later be represented by an AC equivalent network.

A technique to perform AC/DC system reliability analysis in a composite system where a TCSC [19] is employed in the AC link to adjust the transmission in-feed impedance which increase the transmission system capacity. Reliability models associated with the bipolar DC link and the TCSC are developed and illustrated by Firuzabad et. al [46]. Chen and Thorp [52] developed a hidden failure simulation model for the purpose of evaluating power transmission system reliability. The model uses DC load flow approximation and the linear programming technique to simulate cascading blackouts. In reliability evaluation of generating units and transmission lines together computation of accurate estimates of actual frequency, duration and frequency related indices is a difficult problem as it involves recognition of all the load curtailment states that can be reached from a failure state in one transition. The tools to evaluate composite generation and transmission systems will have to deal with very large power networks.

The methodology of reliability evaluation of the transmission system is based on the concept that the reliability level of a transmission system is equal to the difference in the reliability levels of the composite power system (HL-II) and generation system (HL-I) [31]. A new approach for evaluating the health of composite generation and transmission system was presented by DaSilva et. al [60]. A well-being framework was used to clarify the system states.

The most common application of FACTS devices as well as their benefits was defined by Habur and O'Leary [70]. Generic information on the costs and benefits of FACTS devices is then provided as well as the steps for identification of FACTS projects. The utilization of FACTS technologies can have sufficient positive impacts on power system reliability performance and the actual benefits obtained can be quantitatively assessed using suitable models and techniques were stated [50]. An optimization model to be used in composite power system reliability evaluation method when employing a Static Synchronous Compensator is stated by Bay and Kazemi [73]. Some probabilistic load point indices and system indices have been calculated in their paper.

Flexible AC Transmission System (FACTS) technology is the ultimate tool for getting the most out of existing equipment viz. faster control action and new capabilities. FACTS technology is a tool for permitting existing transmission facilities to be loaded at least under contingency situations, upto their thermal limits without degrading system security. The most striking feature is the ability to directly control transmission line flows by structurally changing parameters of the fast switching.

Billinton et. al [33] described the impact of a TCSC on power system reliability. TCSC is employed to adjust the natural power sharing [75] of two different parallel transmission lines and therefore enable the maximum transmission capacity to be utilized. Thyristor Controlled Series Capacitors which offer a strong alternative for optimizing of transmission over power links, existing as well as new, by means of increased dynamic stability, power oscillation damping as well as optimized load flow between parallel circuits [35]. There are many factors that influence the effects of a TCSC. For instance, the system situation might worsen due to the failure of a TCSC. Different placements of TCSC can have different degrees of impact on the load curtailments then the system reliability. The parameters of the network such as thermal limits also affect the efficiency of TCSC. TCSCs can be employed in a system to adjust the transmission in feed Impedances [63] and therefore increase the transmission system capacity without increasing the system fault current levels [43]. Distribution system capacity additions required to serve load growth are generally accommodated in one of two ways; either additional feeders can be extended, or load can be transferred to a new substation and its associated feeders. Although substation capacity upgrades are the preferable option, these are, in some cases, severely constrained by limits to existing circuit breaker interrupting capability. A reliability model of a multi module TCSC was developed and incorporated in the transmission system [33]. In further investigations [40] the authors have presented an approach to evaluate transmission system reliability when employing a Unified Power Flow Controllers. They [40] provided a framework within which the risk associated with management and the development of the transmission network can be quantified [18].

Billinton et. al [44] described that UPFC is employed in the system to adjust the natural power sharing of two different parallel transmission lines and therefore enable the maximum transmission capacity to be utilized. Huang and Li [51] have described the reliability model of a transmission line with a TCSC which is built and simplified via the state space approach. Optimal power flow including the influence of TCSC is then introduced to decide the load curtailment. The impact of an Interline Power Flow Controller (IPFC) on composite system delivery point and overall system reliability indices was examined by Moghadasi et. al [71]. In this application, the IPFC with DC transmission lines is employed in a system for coordinated control of line impedances between transmission lines with the objective of managing power flows on the two lines.

Ou and Singh [47] presented the formulation and general procedure of Total Transfer Capacity (TTC) calculation using TCSC and Static Voltage Compensator (SVC). Improvement of TTC using TCSC and SVC is demonstrated with a simple test system. Both thermal limit dominant case and voltage limit dominant case are investigated. An optimization model to use in composite power system reliability evaluation method incorporating the impact of FACTS devices [75] was proposed by Verma et. al [61]. The conventional DC flow-based linear programming model used in composite system reliability evaluation method is converted into a non-linear optimization model to include the impact of FACTS devices on reliability of power systems. The purpose of assessing the reliability of a composite power generation and transmission system is to estimate the ability of the system to perform its function of moving the energy provided by the generating system to the bulk supply points. Reliability evaluation of composite generation and transmission systems is an important area of concern for system planners and operators.

Some of the methodologies and computational techniques used for composite generation / transmission reliability evaluation which include conceptual framework for reliability evaluation, characterization and selection of system states, sensitivity analysis and techniques for reducing computational effort are surveyed by Pereira and Balu [16]. Appropriate reliability indices and the means of evaluation of these indices is one of the basic procedures in power systems reliability education. The reliability indices [76] are usually based on the continuity of service to the consumer, the unreliability effects on losses which are identified by Bian et. al [17]. The fundamental differences between the techniques for composite system reliability evaluation were discussed, a simulation procedure for bulk power system reliability evaluation in which operation of the systems under forced or scheduled outages of circuits and/or units is simulated via a stochastic load flow are illustrated [11].

A methodology for frequency and duration assessment in composite generation and transmission reliability evaluation in which the approach uses the concept of conditional probability to characterize the contribution of each component to the frequency indices were described by Melo et. al [20]. Latter, Weber et. al [25] reviewed published literature and other documentation of existing procedures, definitions and indices. The report presented the results of the Task force efforts to compile a useful set of terms and procedures for consistent reporting of Bulk Power Supply (BPS) delivery point reliability [4]. The effects of bus load uncertain and correlation in composite system adequacy assessment in which tabulating techniques of normal distribution sampling and a correlation sampling technique are utilized to simulate bus load uncertainty and correlation were illustrated by Wenyan and Billinton [15]. Toscano et. al [72] presented a new dynamic reliability approach which is able to take into account the various operating conditions. Billinton and Adzanu introduced [29] the development of an annual chronological load curve for each load bus in composite generation and transmission system [77] in which the load curves are then utilized in a composite system adequacy assessment using the analytical or contingency enumeration approach.

Celo and Bualoti [65] analyzed the reliability indices and the factors that have significant influence on the system reliability indices in which failure of generating units and transmission lines are considered where station components failure results in line outages are accounted for adjustments in the line failure rates. A new analytical way to calculate Expected Energy Not Supplied (EENS) is proposed by Zhu [67].

From the literature so far available, FACTS components like TCSC, UPFC and IPFC have been incorporated independently in the transmission lines to determine the reliability analysis of the system. The analysis has been performed by considering only the state space representation. No attempt has been made to determine the reliability analysis by combining the FACTS controllers. System indices have not been evaluated by considering the generation data and FACTS controllers simultaneously, in so far available literature.

1.3 Objective:

In this book an attempt, has made to determine the reliability, system indices, probability of failure and EENS by combining the FACTS controllers together for natural sharing of power. Here, the analysis is made by changing both generation and load simultaneously. The analysis is also carried out by using network reduction techniques which is not available in the literature so far. An attempt has been made to develop state space representation when combining two FACTS devices. Comparison of different FACTS devices in a system has been extended from the literature so far available.

- To evaluate reliability analysis of a transmission system which uses FACTS components like TCSC and UPFC in combination with Series compensator. Here the loss indices like Loss of Load Expectation (LOLE), Loss of Energy Expectation (LOEE) will be evaluated. Reliability analysis is to be carried out by using two different techniques like State – Space technique and Series – Parallel representation.
- Proposed to implement FACTS (TCSC and UPFC) components into 6 bus RBTS which is categorized as composite power system. System Indices like Bulk Power Supply Disturbance (BPSD), Bulk Power Interruption Index (BPII), Bulk Power Energy Curtailment Index (BPECI), Probability of failure and Expected Energy Not Supplied (EENS) are being proposed to calculate at each and every bus to determine their respective performance. Above analysis will be carried out by incorporating different modules of FACTS elements in the transmission line. Comparison will be made out for the above analysis with different FACTS elements.
- Proposed to implement the combination of FACTS devices into the 6 Bus RBTS (Roy Billinton Test System) at different location to verify and improve the reliability analysis, System Indices like BPSD (Bulk Power Supply Distrubance), BPII (Bulk Power Interruption Index), BPECI (Bulk Power Energy Curtailment Index), probability of failure and EENS (Expected Energy Not Supplied) at every bus. Reliability analysis of the combination of the FACTS devices is obtained by using State space technique and series-parallel representation.
- It is proposed to determine the system indices and comparison of these will be tested on IEEE 24 Bus RTS.

Chapter 2

RELIABILITY ANALYSIS OF TRANSMISSION SYSTEM USING TCSC AND SERIES COMPENSATOR

2.1 Introduction

The general principle used is to reduce sequentially the complicated configuration by combining appropriate series and parallel branches of the reliability model until a single equivalent element [19] remains. This equivalent element then represents the reliability of the original configuration. If blocks are repeated, an attempt should be made to simplify the structure to eliminate repetition. Not every replications can, however, be eliminated by simplification and it may also happen that a possibly of simplifying the diagram is overlooked. In such cases the dependent relation [8] inherent in replication must be taken into account.

Series compensation with Thyristor Control (TCSC) enables rapid dynamic modulation of the inserted reactance. At interconnection points between transmission grids, this modulation will provide strong damping torque on inter-area electromechanical oscillations. As a consequence, a TCSC makes it possible to interconnect grids having generating capacity in the many thousands of megawatts. Often the TCSC is combined with fixed series compensation to increase transient stability in the most cost effective way.

Series compensation of power transmission circuits enables several useful benefits:
- An increase of active power transmission over the circuit without violating angular or voltage stability;
- An increase of angular and voltage stability without derating power transmission capacity;
- A decrease of transmission losses in many cases.

In this Chapter, the reliability analysis of transmission system using TCSC and Series Compensator is presented. A sample power system is considered for the determination of reliability analysis of a transmission line which is discussed in Section 2.2. The reliability analysis for the combination of TCSC and series compensator can be determined by using either Series-Parallel representation or State Space representation. The analysis of the above combination is discussed by series-parallel representation using network reduction techniques in Section 2.3. However, network reduction techniques cannot be applied for all the systems where the availability of the system should be predicted accurately, state space representation will be used in place of network reduction techniques. The State Space representation of the combination, illustrations is discussed in Section 2.4. Reliability analysis of the proposed transmission system is being carried out by using Load Indices like LOLE & LOEE which is discussed in Section 2.5. Conclusions of Chapter 2 are presented in Section 2.6.

2.2 System under Study

A sample power system is considered as shown in Fig. 2.1. A series compensator is included in one of the lines, say L_1. A series compensator essentially consists of a capacitor bank as shown in Fig. 2.3.

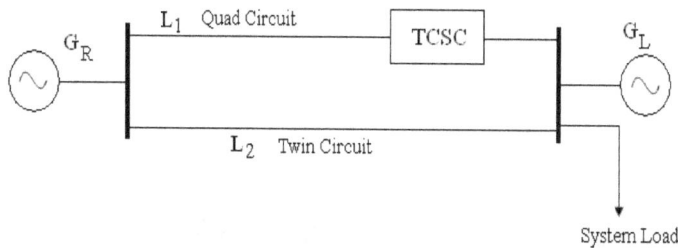

Fig 2.1: System under Study

Specifications of the system under study (Fig.2.1):

It consists of local & remote generating sites interconnected by two transmission line (L_1, L_2) of unequal thermal ratings. The remote generating site provides 41.6% of the total system generation while the local generating site contributed the other 58.4%. Transmission Line L_1 is a four conductor bundle circuit configuration with a rating of 3000MVA, whereas transmission Line L_2 has a two conductor bundle circuit configuration with a rating of 1500MVA. The four conductor bundle circuit is series compensated by TCSC. The degree of compensation is such that the equivalent reactance of the line is one half the inductive reactance of the two conductor bundle circuit.

Table 2.1: Generation Data for the Proposed System

Generation	No. of Units	Capacity (MVA)	Failure Rate (f/yr)	Repair Time (Hrs)
Remote	4	750	5.8	80
	3	350	7.62	100
	2	225	8.5	78
Local	12	375	7.62	100
	4	180	10	98
	5	220	7.5	74

Table 2.2: Transmission Data for the Proposed System

Transmission	Capacity MVA	Line Impedance pu/km (100MVA, 500KV Base)	Failure Rate (f/yr)	Repair Time (hrs)
L_1	3000	0.0012+j0.016	1.5	15
L_2	1500	0.0024+j0.019	0.7	10

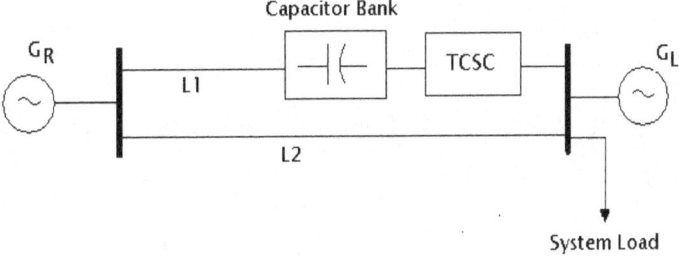

Fig 2.2: Proposed System for Study

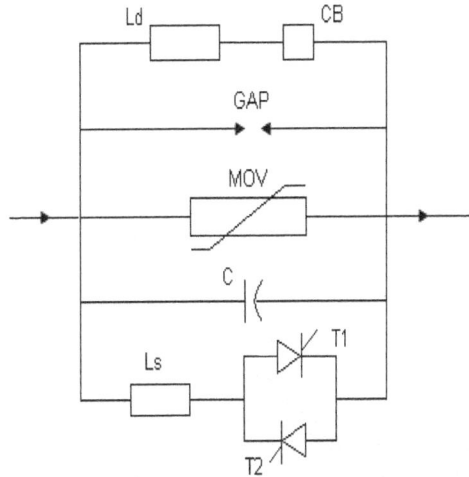

Fig.2.3: Block Diagram of Practical TCSC showing various components

Fig. 2.3 shows a practical TCSC module [1] with different protection elements. Basically it comprises a series capacitor C, in parallel with a Thyristor Controlled Reactor (TCR) Ls. A Metal Oxide Varistor (MOV) essentially a nonlinear resistor is connected across the series capacitor to prevent the occurrence of high capacitor over voltage. Not only does the MOV limit in the circuit even during fault conditions and help improve the transient stability. A circuit breaker is also installed across the TCSC module to bypass it if a severe fault or equipment malfunction [1] occurs. A current limiting inductor, L_d is incorporated in the circuit to restrict both the magnitude and the frequency of the capacitor current during the capacitor bypass operation. It consists of series compensating capacitor shunted by a thyristor controlled reactor. In a practical TCSC implementation, several such basic compensations may be connected in series to obtain the desired voltage rating and operating characteristics. Basic idea behind the TCSC scheme is to provide a continuously variable capacitor by means of partially canceling the effective compensating capacitance by the TCR. The individual reliability indices are taken from the reference [6].

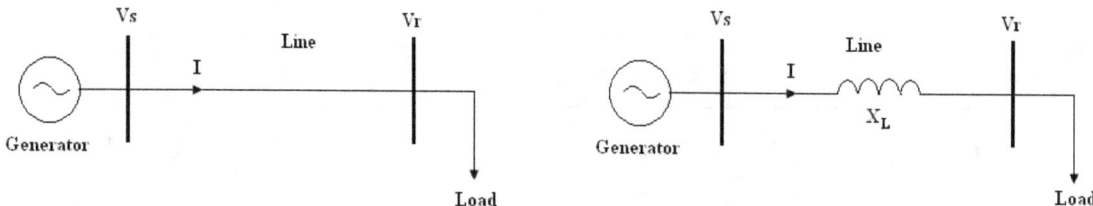

Fig. 2.4: Sample Power System NetworkFig. 2.4(a): Equivalent Circuit of system under study

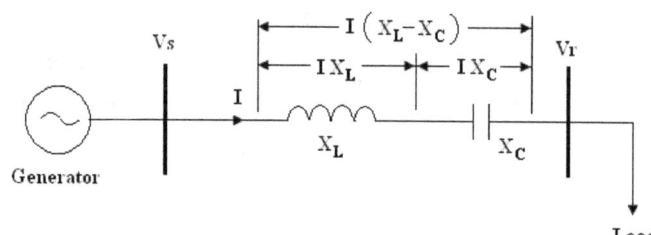

Fig. 2.4(b): Equivalent Circuit after introducing capacitor

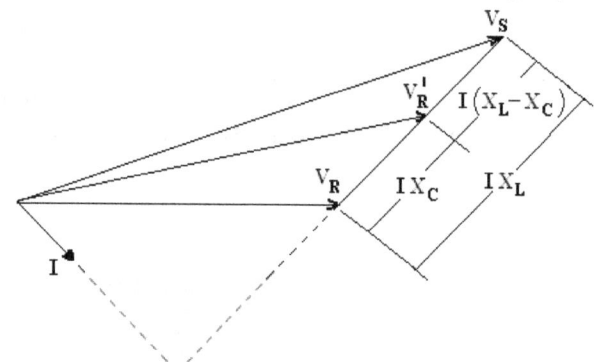

Fig. 2.4(c): Phasor Diagram for Fig. 2.4(b)

2.3 Reliability Logic Diagram Using Series – Parallel System

The Reliability Logic Diagram (RLD) of Thyristor Controlled Series Compensator and Series Compensator using Series – Parallel system is shown in Fig. 2.5. Each rectangle block in the figure represents a particular component. Here each component has its own reliability which is independent of the time. Considering these reliabilities, in combination of simple series and parallel system, the overall reliability and unreliability of the system are determined as follows:

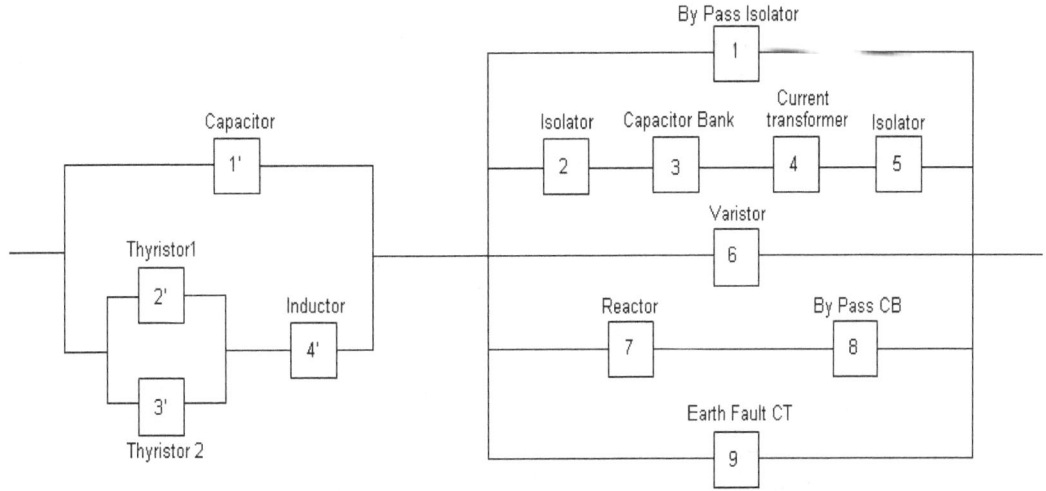

Fig. 2.5: RLD for combination of TCSC and Series Compensator using Series – Parallel System

The series parallel network reductions of TCSC & Series Compensator are shown in Figs. 2.5 (a) to (d) respectively.

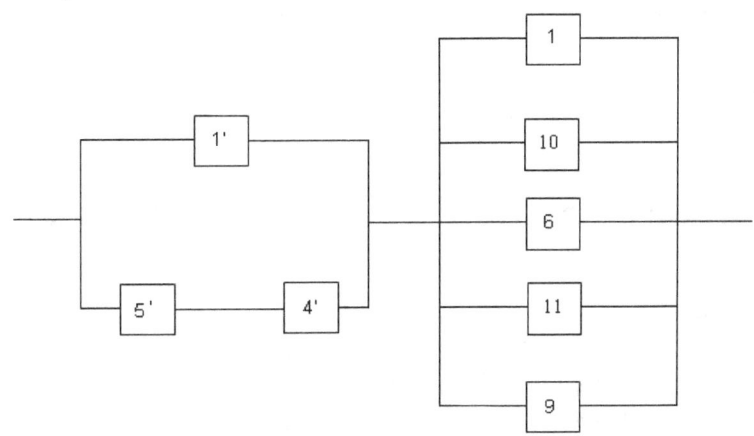

Fig. 2.5 (a): RLD Network Reductions – Step 1

From the Fig. 2.5(a) block
   1' indicates Capacitor in TCSC
   5' indicates the network reduction of 2' & 3' which are in parallel combination
        2' & 3' represents Thyristors of TCSC which are anti-parallel to each other.
   As the block 5' is parallel combination the Reliability of this block can be determined as
R = 1-Q
   4' indicated Inductor
   1 indicates By pass Isolator
   10 indicates the network reduction of 2, 3, 4 & 5 which are in series combination
      2 & 5 indicates Isolator
      3 indicates capacitor bank
      4 indicates current transformer
  6 indicated Varistor
   11 indicate network reduction of 7 & 8 which are in series combination
      7 indicates reactor
      8 indicates Bypass Circuit Breaker
     9 indicates Earth fault Current Transformer

$R_5^1 = 1 - (Q_2^1 Q_3^1)$ where $Q_2^1 = 1 - R_2^1$, $Q_3^1 = 1 - R_3^1$
$R_{10} = R_2 R_3 R_4 R_5$ $\qquad R_{11} = R_7 R_8$ . . . (2.1)

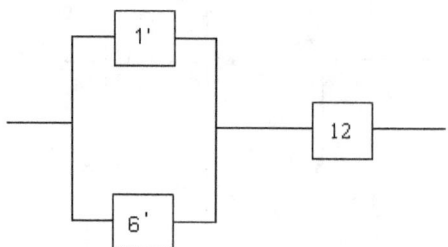

Fig. 2.5 (b): RLD Network Reductions – Step 2

From the Fig. 2.5(b) block
    12 indicate network reduction of 1, 10, 6, 11 & 9 which are in parallel combination
    6' indicates the network reduction of 4' & 5' which are in series combination
    As the block 6' is series combination the Reliability of this block can be determined as
$R_{6'} = R_{5'} R_{4'}$
$R_6^1 = R_5^1 R_4^1$
$R_{12} = 1 - [(1 - R_1)(1 - R_{10})(1 - R_6)(1 - R_{11})(1 - R_9)]$ . . . (2.2)
$Q_{10} = 1 - R_{10}$ $\qquad Q_{11} = 1 - R_{11}$ $\qquad Q_{12} = 1 - R_{12}$

| 7' | 12 | | 13 |

Fig. 2.5 (c): RLD NR – Step 3   **Fig. 2.5 (d): RLD NR – Step 4**

From the Fig. 2.5(c) block
    7' indicates the network reduction of 1' & 6' which are in parallel combination
$$R_7^1 = 1 - (Q_1^1 Q_6^1)$$ . . . (2.3)
    Where $Q_1^1 = 1 - R_1^1$ $\qquad Q_6^1 = 1 - R_6^1$
From the Fig. 2.5(c) block
    13 indicates the network reduction of 7' & 12 which are in series combination
$$R_{13} = R_7^1 R_{12}$$ . . . (2.4)
Where R_7^1 is the reliability of TCSC, R_{12} is the reliability of Series Compensator and R_{13} is the reliability for the combination of TCSC and Series Compensator. 'Q' represents the unreliability of the particular component or system.

2.3.1 Results

Now consider the individual reliabilities of each component [6]:
Bypass Isolator (R_1) = 0.92   Isolator ($R_2 = R_5$) = 0.8
Capacitor Bank (R_3) = 0.85   Current Transformer (R_4) = 0.87
Varistor (R_6) = 0.96   Reactor (R_7) = 0.88
Bypass CB (R_8) = 0.84   Earth Fault CT (R_9) = 0.82
Capacitor [TCSC] (R_1^1) = 0.85   Thyristor ($R_2^1 = R_3^1$) = 0.78
Inductor (R_4^1) = 0.88

Substituting all the reliability values of the components in the above said equations, unreliability / reliability of the overall system is:
    Reliability = 0.9751824
    Unreliability = 0.0248176

2.4 RLD using State Space representation

Reliability Logic Diagram (RLD) is also referred as Reliability Block Diagram (RBD) which performs the system reliability & unavailability analysis on large & complex system using block diagram to show network relationships. RLD is a drawing & calculation tool used to model complex systems. Once the blocks are configured properly & data is provided the failure rate, MTBF, reliability & availability of the system can be calculated. Reliability Logic Diagram is connected by a parallel (or) series configurations.

Series Connection:

A Series connection is joined by the continuous link from the start node to the end node which is shown in Fig.2.6.

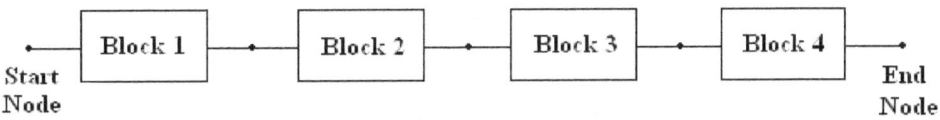

Fig.2.6: Reliability Logic Diagram of Series Connected System

Parallel Connection:

A Parallel connection is used to show redundancy & is joined by multiple links or paths from the Start node to the End Node which is shown in Fig.2.7.

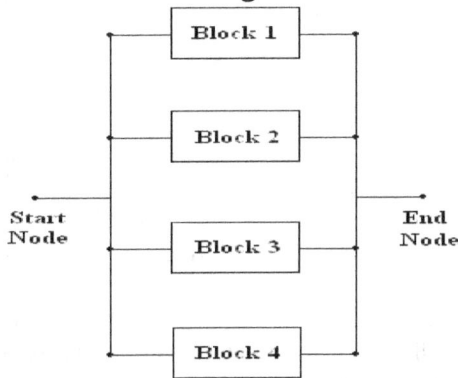

Fig.2.7: Reliability Logic Diagram of Parallel Connected System

A system can contain a series, parallel or combination of Series & Parallel Connections to make up the network which is shown in Fig.2.8. Reliability Block Diagram often corresponds to the physical arrangement of components in the system. However, in certain cases this may not apply.

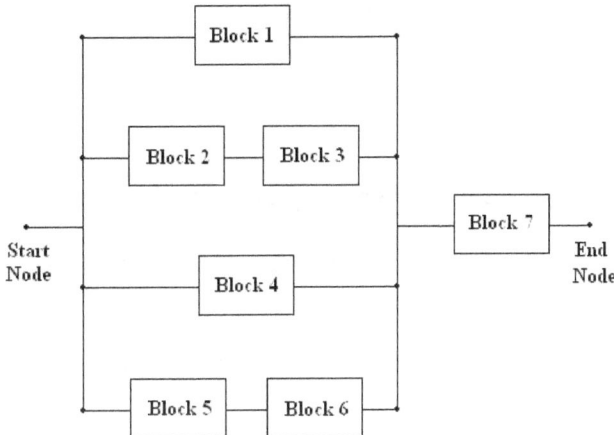

Fig.2.8: Reliability Logic Diagram of Series-Parallel Connected System

The State Space representation for the combination of TCSC and Series Compensator is shown in Fig. 2.10. This is another method for finding the reliability of entire system. Here the rectangular blocks from 1 to 16 represents the transition state out of 19 states with combination of TCSC and series compensator. The upper transition represents the states of TCSC and lower transitions represent states of Series Compensator [19]. Here, only 16 states are considered for the combination of both elements and the remaining is not considered because, the remaining states cannot with stand the rated capacity of the transmission line.

The reliability logic diagram consists of two spares one of TCSC and the other of Series Compensator, because, each state is a combination of these two elements [51]. Each and every state is connected to bypass module, because, at any state the system can be failed due to any faults or improper firing of thyristors or failure of capacitors in series compensator. Each state is assigned with a proper transition number in a sequential manner so that, each state follows the previous sates. A bypass block is also considered because, to reduce the capacitor voltage which is due to fault currents.

In order to facilitate the solution of continuous processes it is desirable first to construct the appropriate state space diagram and insert the relevant transition rates. All relevant transition states in which the system can reside should be included in such a diagram and all known ways in which transitions between states can occur should be inserted. There are no basic restrictions on the number of states or the type and number of transitions that can be inserted.

The analyst must therefore first translate the operation of the system into a state space diagram recognizing both the states of the system, the way these states communicate and the value of the transition rates. It is relatively easy to formulate state space diagrams for small system models. Although this method is accurate, it becomes infeasible for large distribution networks. However it has an important role to play in power system reliability evaluation. Firstly, it can be used as the primary evaluation method in certain applications. Secondly, it is frequently used as a means of deducing approximate evaluation technique. Thirdly, it is extremely useful as a standard evaluation method against which the accuracy of approximate methods can be compared.

If the system consists of two repairable components, there are four possible states in which the system can exist. If λ_1, μ_1 and λ_2, μ_2 are the failure and repair rates of components 1 and 2 respectively, the state space diagram including the relevant transition rates is shown in Fig. 2.9.

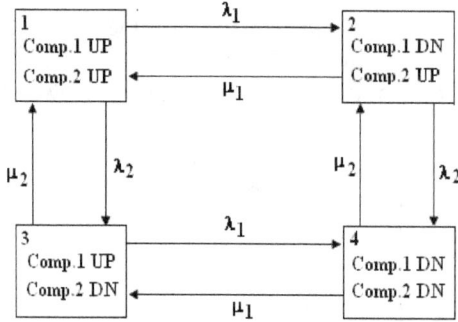

Fig. 2.9 State Space Diagram for two component system.

In the case of a series system state 1 is the system UP state and state 2, 3 and 4 are the DOWN states. In case of parallel redundant system states 1, 2, 3 are the system UP states and state 4 is the system DOWN states.

Therefore if P1, P2, P3 and P4 are the probabilities of being in states 1-4 respectively, then

For the series system $P_{UP} = A = P_1$

$$P_{DOWN} = U = P_2 + P_3 + P_4 \qquad \text{...... (1)}$$

For the parallel redundant system $P_{UP} = A = P_1 + P_2 + P_3$

$$P_{DOWN} = U = P_4 \quad \ldots\ldots (2)$$

And these equations apply for both time dependent and limiting state probabilities. The advantage of the state space method is therefore independent of the configuration but becomes larger depending upon the number of components.

Each state is shown by a rectangle in which the enclosed left side numbers are associated respectively with the state number. The capacities associated with state 2 to 16 are proportional to the number of the available modules. The emergency state or spare state, state 17 and 18, is a transition state between states 1, 2, 3, 4, 5, 6, 7, 8, 9, 10, 11, 12, 13, 14, 15, 16 and state 19.

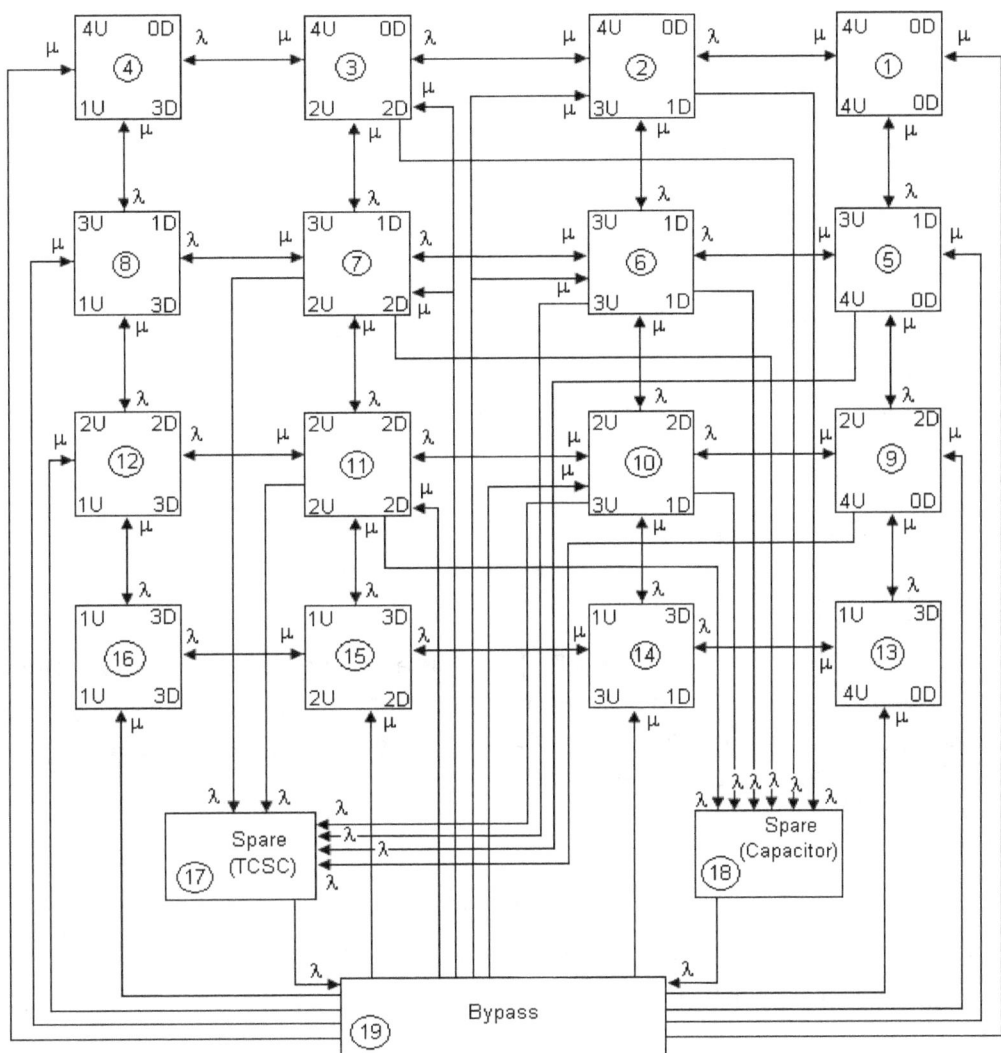

Fig. 2.10: RLD for combination of TCSC and Series Compensator using State – Space Representation

From the above reliability logic diagram, matrix P, which is a Stochastic Transitional Probability Matrix (STPM) is defined as:

$$P = \begin{bmatrix}
1-2\lambda & \lambda & 0 & 0 & \lambda & 0 & 0 & 0 & 0 & 0 & 0 & 0 & 0 & 0 & 0 & 0 & 0 & 0 & 0 \\
\mu & 1-(\mu+3\lambda) & \lambda & 0 & 0 & \lambda & 0 & 0 & 0 & 0 & 0 & 0 & 0 & 0 & 0 & 0 & 0 & \lambda & 0 \\
0 & \mu & 1-(\mu+3\lambda) & \lambda & 0 & 0 & \lambda & 0 & 0 & 0 & 0 & 0 & 0 & 0 & 0 & 0 & 0 & \lambda & 0 \\
0 & 0 & \mu & 1-(\mu+\lambda) & 0 & 0 & 0 & \lambda & 0 & 0 & 0 & 0 & 0 & 0 & 0 & 0 & 0 & 0 & 0 \\
\mu & 0 & 0 & 0 & 1-(\mu+3\lambda) & \lambda & 0 & 0 & \lambda & 0 & 0 & 0 & 0 & 0 & 0 & 0 & \lambda & 0 & 0 \\
0 & \mu & 0 & 0 & \mu & 1-(2\mu+4\lambda) & \lambda & 0 & 0 & \lambda & 0 & 0 & 0 & 0 & 0 & \lambda & \lambda & 0 & 0 \\
0 & 0 & \mu & 0 & 0 & \mu & 1-(2\mu+4\lambda) & \lambda & 0 & 0 & \lambda & 0 & 0 & 0 & 0 & \lambda & \lambda & 0 & 0 \\
0 & 0 & 0 & \mu & 0 & 0 & \mu & 1-(2\mu+\lambda) & 0 & 0 & 0 & \lambda & 0 & 0 & 0 & 0 & 0 & 0 & 0 \\
0 & 0 & 0 & 0 & \mu & 0 & 0 & 0 & 1-(\mu+3\lambda) & \lambda & 0 & 0 & \lambda & 0 & 0 & 0 & \lambda & 0 & 0 \\
0 & 0 & 0 & 0 & 0 & \mu & 0 & 0 & \mu & 1-(2\mu+4\lambda) & \lambda & 0 & 0 & \lambda & 0 & 0 & \lambda & \lambda & 0 \\
0 & 0 & 0 & 0 & 0 & 0 & \mu & 0 & 0 & \mu & 1-(2\mu+4\lambda) & \lambda & 0 & 0 & \lambda & 0 & \lambda & \lambda & 0 \\
0 & 0 & 0 & 0 & 0 & 0 & 0 & \mu & 0 & 0 & \mu & 1-(2\mu+\lambda) & 0 & 0 & 0 & \lambda & 0 & 0 & 0 \\
0 & 0 & 0 & 0 & 0 & 0 & 0 & 0 & \mu & 0 & 0 & 0 & 1-(\mu+\lambda) & \lambda & 0 & 0 & 0 & 0 & 0 \\
0 & 0 & 0 & 0 & 0 & 0 & 0 & 0 & 0 & \mu & 0 & 0 & \mu & 1-(2\mu+\lambda) & \lambda & 0 & 0 & 0 & 0 \\
0 & 0 & 0 & 0 & 0 & 0 & 0 & 0 & 0 & 0 & \mu & 0 & 0 & \mu & 1-(2\mu+\lambda) & \lambda & 0 & 0 & 0 \\
0 & 0 & 0 & 0 & 0 & 0 & 0 & 0 & 0 & 0 & 0 & \mu & 0 & 0 & \mu & 1-2\mu & 0 & 0 & 0 \\
0 & 0 & 0 & 0 & 0 & 0 & 0 & 0 & 0 & 0 & 0 & 0 & 0 & 0 & 0 & 0 & 1-\lambda & 0 & \lambda \\
0 & 0 & 0 & 0 & 0 & 0 & 0 & 0 & 0 & 0 & 0 & 0 & 0 & 0 & 0 & 0 & 0 & 1-\lambda & \lambda \\
\mu & \mu & \mu & \mu & \mu & \mu & \mu & \mu & \mu & \mu & \mu & \mu & \mu & \mu & \mu & \mu & 0 & 0 & 1-16\mu
\end{bmatrix}$$

$[P_{SS}][P] = [P_{SS}]$

Where, $P_{SS} = [P_1\ P_2\ P_3 \ldots\ldots\ldots\ldots P_{17}\ P_{18}\ P_{19}]$ which is a limiting state probability vector.

The following equations are developed from the expression $[P_{SS}] \times [P] = [P_{SS}]$. The above expression is derived by multiplying with P_{SS} which is a limiting state probability vector both sides of STPM (Stochastic Transitional Probability Matrix).

Expressing the above matrix form in terms of equations,

$$P_1(1-2\lambda) + P_2\mu + P_5\mu + P_{19}\mu = P_1 \quad \cdots \quad (2.5)$$

$$P_1\lambda + P_2(1-(\mu+3\lambda)) + P_3\mu + P_6\mu + P_{19}\mu = P_2 \quad \cdots \quad (2.6)$$

$$P_2\lambda + P_3(1-(\mu+3\lambda)) + P_4\mu + P_7\mu + P_{19}\mu = P_3 \quad \cdots \quad (2.7)$$

$$P_3\lambda + P_4(1-(\mu+\lambda)) + P_8\mu + P_{19}\mu = P_4 \quad \cdots \quad (2.8)$$

$$P_1\lambda + P_5(1-(\mu+3\lambda)) + P_6\mu + P_9\mu + P_{19}\mu = P_5 \quad \cdots \quad (2.9)$$

$$P_2\lambda + P_5\lambda + P_6(1-(2\mu+4\lambda)) + P_7\mu + P_{10}\mu + P_{19}\mu = P_6 \quad \cdots \quad (2.10)$$

$$P_3\lambda + P_6\lambda + P_7(1-(2\mu+4\lambda)) + P_8\mu + P_{11}\mu + P_{19}\mu = P_7 \quad \cdots \quad (2.11)$$

$$P_4\lambda + P_7\lambda + P_8(1-(2\mu+\lambda)) + P_{12}\mu + P_{19}\mu = P_8 \quad \cdots \quad (2.12)$$

$$P_5\lambda + P_9(1-(\mu+3\lambda)) + P_{10}\mu + P_{13}\mu + P_{19}\mu = P_9 \quad \cdots \quad (2.13)$$

$$P_6\lambda + P_9\lambda + P_{10}(1-(2\mu+4\lambda)) + P_{11}\mu + P_{14}\mu + P_{19}\mu = P_{10} \quad \cdots \quad (2.14)$$

$$P_7\lambda + P_{10}\lambda + P_{11}(1-(2\mu+4\lambda)) + P_{12}\mu + P_{15}\mu + P_{19}\mu = P_{11} \quad \cdots \quad (2.15)$$

$$P_8\lambda + P_{11}\lambda + P_{12}(1-(2\mu+\lambda)) + P_{16}\mu + P_{19}\mu = P_{12} \quad \cdots \quad (2.16)$$

$$P_9\lambda + P_{13}(1-(\mu+\lambda)) + P_{14}\mu + P_{19}\mu = P_{13} \quad \ldots \quad (2.17)$$

$$P_{10}\lambda + P_{13}\lambda + P_{14}(1-(2\mu+\lambda)) + P_{15}\mu + P_{19}\mu = P_{14} \quad \ldots \quad (2.18)$$

$$P_{11}\lambda + P_{14}\lambda + P_{15}(1-(2\mu+\lambda)) + P_{16}\mu + P_{19}\mu = P_{15} \quad \ldots \quad (2.19)$$

$$P_{12}\lambda + P_{15}\lambda + P_{16}(1-2\mu) + P_{19}\mu = P_{16} \quad \ldots \quad (2.20)$$

$$P_5\lambda + P_6\lambda + P_7\lambda + P_9\lambda + P_{10}\lambda + P_{11}\lambda + P_{17}(1-\lambda) = P_{17} \quad \ldots \quad (2.21)$$

$$P_2\lambda + P_3\lambda + P_6\lambda + P_7\lambda + P_{10}\lambda + P_{11}\lambda + P_{18}(1-\lambda) = P_{18} \quad \ldots \quad (2.22)$$

$$P_{17}\lambda + P_{18}\lambda + P_{19}(1-16\mu) = P_{19} \quad \ldots \quad (2.23)$$

Since all the above Eqns. (2.5 to 2.23) are independent to each other, we consider only 18 equations out of the above 19 equations and 19th equation is taken as

$$P_1+P_2+P_3+P_4+\ldots\ldots\ldots\ldots+P_{17}+P_{18}+P_{19} = 1 \quad \ldots \quad (2.24)$$

Writing the above Eqns. (2.4 to 2.22) and (2.24) in matrix form,

$$\ldots \quad (2.25)$$

2.4.1 Results

From the above matrix form (Eqn. 2.25), find the limiting state probabilities. Consider the data: Failure Rate (λ) = 0.7 f/yr

Repair Rate (μ) = 150 hrs of each component, then

Individual Limiting State Probabilities are:

P_1 = 0.9814	P_2 = 0.0046	P_3 = 0.00025	P_4 = 1.2 e-4
P_5 = 2.3 e-5	P_6 = 11.2 e-6	P_7 = 23.2 e-6	P_8 = 13.2 e-7
P_9 = 43.8 e-8	P_{10} = 47.8 e-8	P_{11} = 12.8 e-9	P_{12} = 53.2 e-10
P_{13} = 12.6 e-11	P_{14} = 11.8 e-12	P_{15} = 14.7 e-13	P_{16} = 12.3 e-14
P_{17} = 0.0068	P_{18} = 0.0068	P_{19} = 27.3 e-6	

Therefore, the sum of the limiting state probabilities is
$P_1+P_2+P_3+P_4+P_5+P_6+P_7+P_8+P_9+P_{10}+P_{11}+P_{12}+P_{13}+P_{14}+P_{15}+P_{16}+P_{17}+P_{18}+P_{19} = 1$

Availability of the system (P_{UP}) = $P_1+P_{17}+P_{18}$ = 0.9814 + 0.0068 + 0.0068 = **0.995**
Unavailability (P_{DOWN}) = 1-0.995 = **0.005**

2.5 Load Indices

Reliability analysis of the entire transmission system is being carried out by using load indices [61] like Loss of Load Expectation (LOLE) and Loss of Energy Expectation (LOEE) [11].

The variations in the LOEE and LOLE [8] with system peak load are shown in Figs. 2.11 & 2.12 respectively. The annual load factor is assumed to be 70%. It can be seen that, for a give peak load, the LOEE and LOLE decrease with the employment of the TCSC. The inclusion of the TCSC allows the capacity of transmission line to be extended to its thermal limit and therefore, transfer more available capacity from the remote generation site to the load point. The inclusion of series compensator allows reducing the effect of inductance.

2.5.1 Calculations for LOEE:

LOEE (Loss of Energy Expectation) is the expected energy that will not be supplied due to those occasions when the load demand exceeds the available capacity. The probabilities of having varying amounts of capacity unavailable are combined with the system load. Any outage of generating capacity exceeding the reserve will result in a curtailment of system load energy.

$$LOEE = \sum_{i=1}^{n} E_k P_k \quad \text{....... (3)}$$

Where O_k = Magnitude of the Capacity Outage
P_k = Probability of a Capacity Outage equal to O_k
E_k = Energy Curtailed by a capacity outage equal to O_k

The probable energy curtailed is $E_k P_k$. The sum of these products is the total expected energy curtailment or loss of energy expectation (LOEE).
For Remote Generation, U = 0.01 & A = 0.99

Table 2.3: EENS for Remote Generation

Capacity out of Service	Capacity in Service	Probability
0	5880	0.922
735	5145	0.07456
1470	4410	2.6348×10^{-3}
2205	3675	5.325×10^{-5}
2940	2940	6.7235×10^{-7}
3675	2205	5.43312×10^{-9}
4410	1470	2.74428×10^{-11}
5145	735	7.92×10^{-14}
5880	0	1×10^{-6}

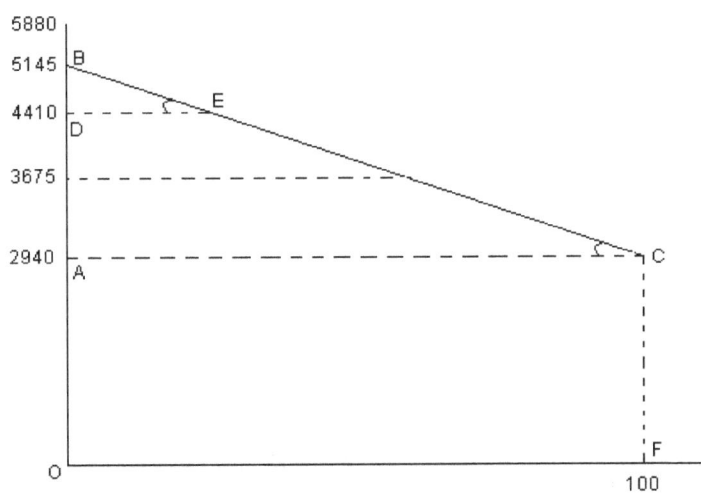

Individual probabilities can be found using the relation $^nC_r\,R^{n-r}\,Q^r$
From the Fig. $\tan\theta = (5145 - 2940) / 100 = 22.05$, $\theta = \tan^{-1}(22.05) = 87.403$

Total energy curtailed = Total area of curve
= Area of OACF + Area of ACB
= 2940*100 + 0.5 * 100 * 2205 = 404.25 GW

For 735 MW capacity outage = 0.5 * 4410 * 200 = 441 GW = 10.451 GW hr/yr
Therefore, the expected energy not supplied for remote generation is:
EENS (GW hr) = 10.451 + 3.6 + 1.5 + 1.0 + 0.85 = 17.401 GW hr/yr

For Local Generation (1),

Table 2.4: EENS for Local Generation (1)

Capacity out of Service	Capacity in Service	Probability
0	2060	0.9801
1030	1030	0.0198
2060	0	0.0001

Therefore, the expected energy not supplied for local generation (1) is:
EENS (GW hr) = 3.97 + 2.11 + 0.89 = 6.97 GW hr/yr

For Local Generation (2),

Table 2.5: EENS for Local Generation (2)

Capacity out of Service	Capacity in Service	Probability
0	560	0.9801
280	280	0.0198
560	0	0.0001

Therefore, the expected energy not supplied for Local Generation is:
EENS (GW hr) = 2.1 + 1.2 + 0.7 = 4.0 GW hr/yr
Total EENS (Local (1 & 2) + Generation) = 17.401 + 6.97 + 4.0
= 28.371 GW hr/yr

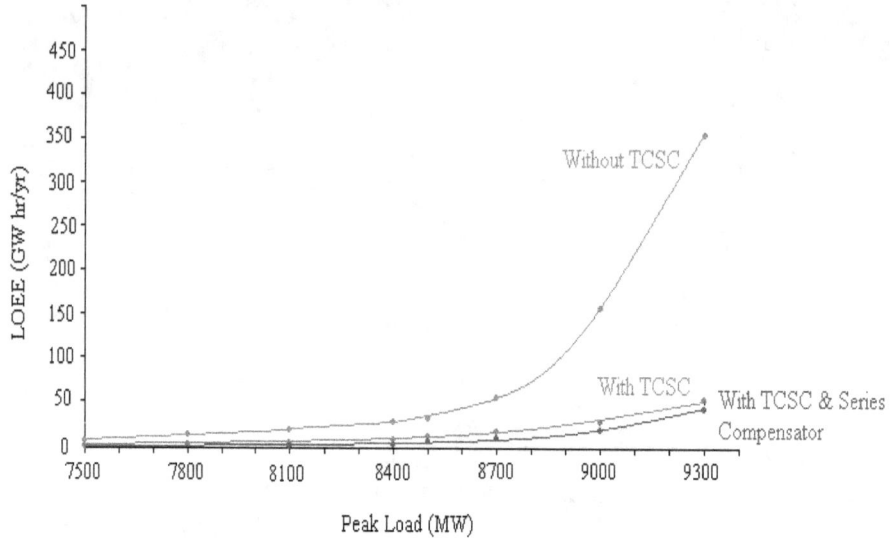

Fig. 2.11: Peak Load (MW) vs LOEE (GWhr/yr) wrt TCSC

Addition of capacitance in series with the transmission line modifies the reactance of the line. The difference is very small for lower system peak loads, as the local generation has sufficient capacity to prevent load interruption. The TCSC effect becomes significant as the peak load increases.

2.5.2 Calculations for LOLE:

Peak Load is considered as 8500 MW, Number of occurrence is 12, 83, 107, 116 and 47 for peak loads of 7940, 7470, 6910, 6440 and 5145 respectively for individual probability of capacity in service.

Table 2.6: Data for Generation System

Generation	No. of Units	Capacity [MW]
Remote	8	735
Local	2	1030
	2	280

Total capacity = 8*735 + 2*1030 + 2*280 = 5880 + 2060 + 560 = 8500MW
Failure rate (λ) = 0.01, Repair rate (μ) = 0.49
Availability (A) = 0.49 / (0.01+0.49) = 0.98
Unavailability (U) = 1 - 0.98 = 0.02
Individual probabilities can be found using the relation $^nC_r\ R^r\ Q^{n-r}$

Table 2.7: Individual Probabilities for Loss of Load Occurrence

Capacity Out of Service	Capacity in Service	Individual Probability
0	8500	0.7847
560	7940	0.192
1030	7470	0.0215
1590	6910	$1.466*10^{-3}$
2060	6440	$6.732*10^{-5}$

2620	5880	2.1998*10⁻⁶
3355	5145	5.238*10⁻⁸
4825	3675	9.163*10⁻¹⁰
5560	2940	1.1687*10⁻¹¹
6295	2205	1.06*10⁻¹³
7030	1470	6.49*10⁻¹⁶
7765	735	2.408*10⁻¹⁸
8500	0	4.096*10⁻²¹

LOLE = 12 P_i(8500-7940) + 83 P_i(8500-7470) + 107 P_i(8500-6910) + 116 P_i(8500-6440) + 47 P_i(8500-5145)

= 12 P_i 560 + 83 P_i 1030 + 107 P_i 1590 + 116 P_i 2060 + 47 P_i 3355

= 12*0.192 + 83*0.0215 + 107*1.46*10⁻³ + 116*6.732*10⁻⁵ + 47*5.23*10⁻⁸

= 4.252 days / yr = 102 hr / yr

LOLE is computed for the system shown in Fig. 2.1. LOLE is computed using the relation $\text{LOLE} = \sum_{i=1}^{n} P_i(C_i - L_i)$ days/period. The system load factor is varied from 50 to 100% in steps of 5% for the analysis.

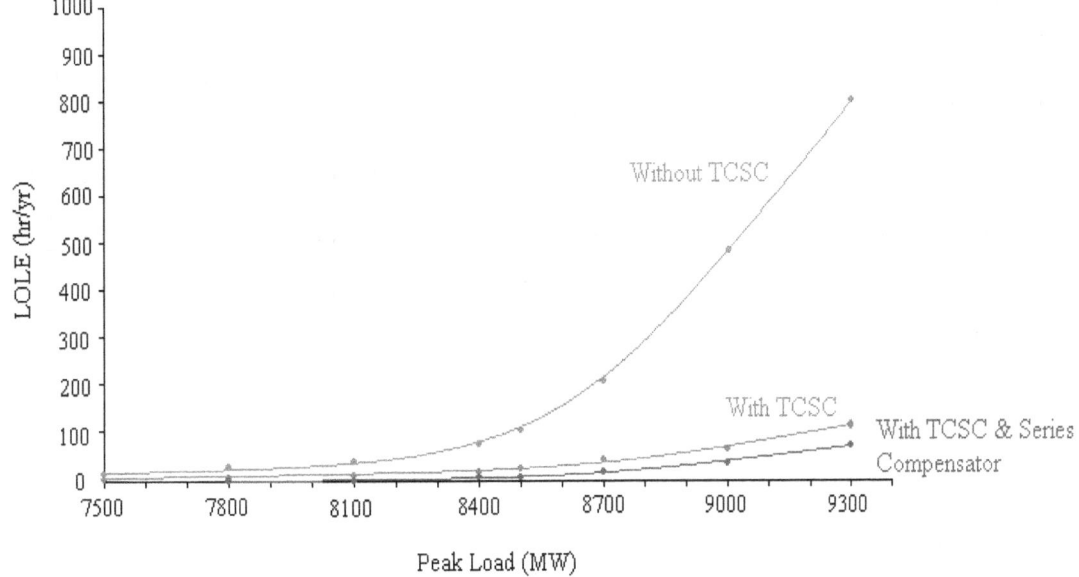

Fig. 2.12: Peak Load (MW) vs LOLE (hr/yr) wrt TCSC

Fig. 2.11 & 2.12 shows the variation in LOLE & LOEE with respect to system peak load respectively. The annual load factor is assumed to be 70%. It can be seen that, for a given peak load, the LOLE & LOEE decreases with the employment of TCSC & Series Compensator. The inclusion of TCSC & Series compensator allows the capacity of transmission line L_1 to be extended to its thermal limits and therefore transfer more available capacity from the remote generation site to the load point. The difference is very small for lower system peak loads, as the local generation has sufficient capacity to prevent load interruption. The TCSC effect becomes significant as the peak load increases.

The impact of TCSC and series compensator on the LOEE and LOLE [8] are shown in Figs. 2.13 & 2.14 respectively. The system load factor is varied from 50% to 100% in steps of 5%. It can be seen from these Figs. 2.13 & 2.14 that the LOEE and LOLE increases with the system load factor.

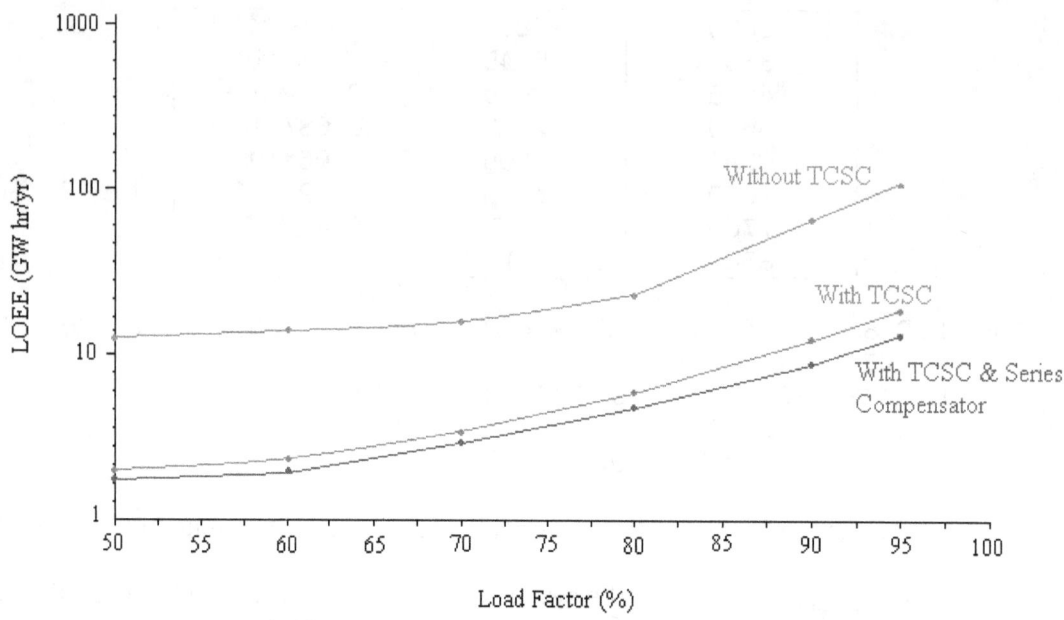

Fig. 2.13: Load Factor (%) vs LOEE (GW hr/yr) wrt TCSC

LOEE when the annual system peak load is assumed to be 8500MW respectively. The system load factor is varied from 50 to 100% in steps of 5%. It can be seen from the figures that LOLE & LOEE increases with increase in the system load factor. It can be seen that for a given load factor, the system reliability is considerably improved by the inclusion of TCSC & Series Compensator.

The results are tabulated in Table 2.8 to 2.11 for different factors for constant system peak load. It can be seen from table that for a given load factor, the system reliability is considerably improved by the inclusion of TCSC and series compensator. Tables 2.8 to 2.11 are developed for the system which is shown in Fig. 2.1.

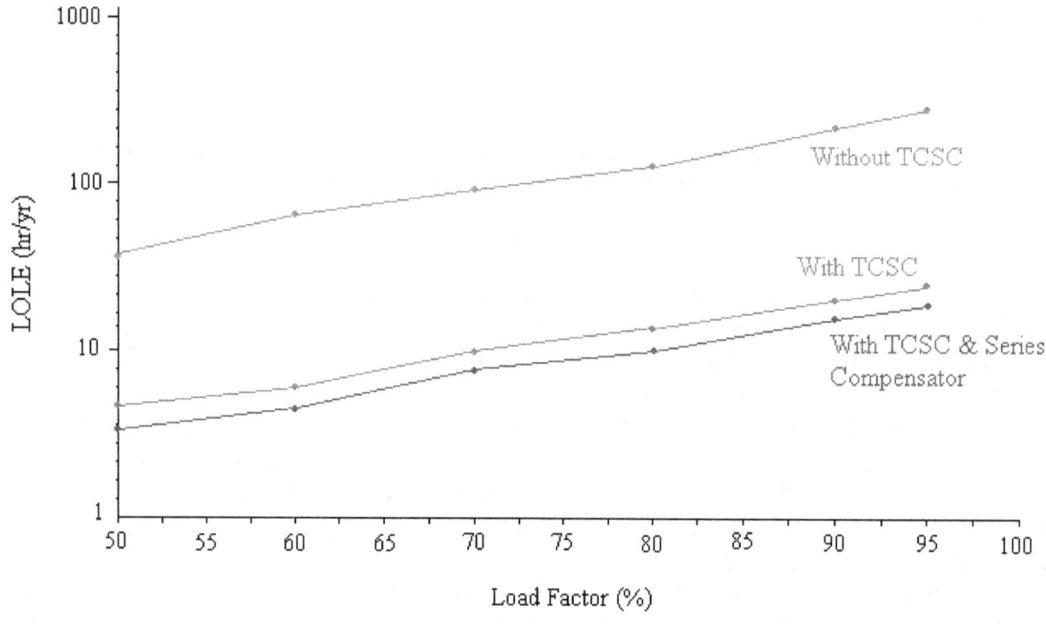

Fig. 2.14: Load Factor (%) vs LOLE (hr/yr) wrt TCSC

Table 2.8: Load (MW) vs LOLE (hr/yr) wrt TCSC

Load MW	LOLE		
	without TCSC	with TCSC	TCSC & SC
7800	15	2	1.8
8100	22	3	2.7
8400	65	8	7.1
8500	101	13	10.7
8700	204	23	19.4
9000	523	54	47.5

Table 2.9: Load (MW) vs LOEE (GW hr/yr) wrt TCSC

Load MW	LOEE		
	without TCSC	with TCSC	TCSC & SC
7500	2	0.1	0
7800	5	0.7	0.52
8100	11.346	1.8	1.3
8400	24.268	3.92	2.97
8500	27.974	5.86	4.17
8700	52.131	10.92	8.62
9000	154.231	23.261	20.16

Table 2.10: Load Factor vs LOLE (hr/yr) wrt TCSC

Load Factor	LOLE		
	without TCSC	with TCSC	TCSC & SC
50	60	7	5.8
60	83	8	6.2
70	101	13	11.7
80	198	19	16.1
90	303	38	33.4
95	401	46	39.2

Table 2.11: Load Factor vs LOEE (GW hr/yr) wrt TCSC

Load Factor	LOEE		
	without TCSC	with TCSC	TCSC & SC
50	18.241	3.624	3.217
60	20.912	4.138	3.838
70	27.974	5.86	5.127
80	41.215	9.631	8.313
90	83.809	17.48	16.006
95	103.713	26.416	24.318

From Tables 2.8 and 2.9 it can be observed that, for any load LOLE & LOEE is decreasing for different configurations viz. without TCSC, with TCSC and TCSC & Series Compensator, which indicates that the combination TCSC & SC is preferable for reduction of losses either in load or in energy indices. Similarly from Tables 2.10 and 2.11, instead of load (MW) load factor of the system is considered to prove the reduction in losses.

2.6 Conclusions

In this chapter, reliability analysis of TCSC and series compensator is estimated with deterministic probability values with the available component data. Further, an attempt also has been made to estimate the availability of the combination of TCSC and series compensator using time dependant probabilities. These time dependant probabilities assume exponential distribution of the failure of the components whereas the deterministic probability data is based on binomial distribution. Load indices are also calculated using load factor & system load for the sample Power Systems Network.

Chapter 3

RELIABILITY ANALYSIS OF TRANSMISSION SYSTEM USING UPFC AND SERIES COMPENSATOR

3.1 Introduction

The general principle used is to reduce sequentially [14] the complicated configuration by combining appropriate series and parallel branches of the reliability model until a single equivalent element remains. This equivalent element than represents the reliability of the original configuration. If blocks are repeated, an attempt should be made to simply the structure to eliminate repetition [18]. Not every replications can, however, be eliminated by simplification and it may also happen that a possibly of simplifying the diagram as it is not possible foe complex systems. In such cases the dependent relation inherent in replication must be taken into account.

For the reliability analysis of the system using UPFC & Series Compensator, the system consider for reliability analysis is shown in Fig. 2.1 and the system description is presented in section 2.2, where, the TCSC is replaced by UPFC. Whatever the analysis performed in determining the reliability indices of a system when using TCSC, is the same, which is applicable when the system is incorporated with UPFC.

In this chapter, the reliability analysis of transmission system using UPFC and Series compensator is presented. The reliability analysis for the combination of UPFC & Series Compensator is discussed by series-parallel representation using network reduction techniques in Section 3.3. However, network reduction techniques cannot be applied for all the systems where the availability of the system should be predicted accurately, state space representation will be used in place of network reduction techniques. The State Space representation of the combination, illustrations is discussed in Section 3.4. Reliability analysis of the proposed transmission system is being carried out by using Load Indices like LOLE & LOEE which is discussed in Section 3.5. Comparison of load indices for both TCSC & UPFC along with series compensator is discussed in Section 3.6. Conclusions of Chapter 3 are presented in Section 3.7.

3.2 Modeling of UPFC

A UPFC consists of two switching converters where each converter is a voltage source inverter using gate turn off thyristor valves as shown in Fig. 3.1. UPFC functions as an ideal AC to AC power converter in which the real power can flow freely in either direction between the AC terminals of the two inverters and each inverter can independently generate reactive power at its own AC output terminals.

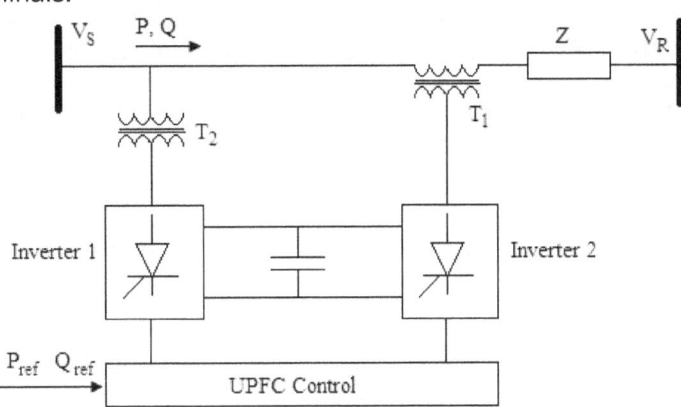

Fig.3.1 Practical Model of UPFC

Inverter 2 provides the main function of the UPFC by injecting an AC voltage V_1 with controllable magnitude & phase angle α at the power frequency, in series with the line can be considered essentially as a synchronous [1] AC voltage source. The real power exchanged at the AC terminal is converted by the inverter into DC power which appears at the DC link as positive & negative real power demand.

The basic function of converter 1 is to supply or absorb the real power demanded by converter 2 at the common DC link to support the real power exchange resulting from the series voltage injection [8]. This DC link power demand of converter 2 is converted back to AC by converter 1 and coupled to the transmission line bus via a shunt connected transformer. Obviously, there can be no reactive power flow through the UPFC DC link [6].

The transmitted power P and the reactive power $-jQ$, supplied by the receiving end, can be expressed [9] as follows:

$$P - jQ = V_r \left(\frac{V_S + V_{pq} - V_r}{jX} \right)^* \quad (3.1)$$

where P = Active Power, Q = Reactive Power
V_r = Receiving end terminal voltage
V_s = Sending end terminal voltage
V_{pq} = Injected voltage
X = Reactance of the line

The transmittable real power P is

$$P_O(\delta) - \frac{VV_{pq\,max}}{X} \leq P_O(\delta) \leq P_O(\delta) + \frac{VV_{pq\,max}}{X} \quad (3.2)$$

Reactive Power Q is

$$Q_O(\delta) - \frac{VV_{pq\,max}}{X} \leq Q_O(\delta) \leq Q_O(\delta) + \frac{VV_{pq\,max}}{X} \quad (3.3)$$

Where δ = Transmission Angle
$P_o(\delta)$ = Normalized transmitted Power
$Q_o(\delta)$ = Normalized reactive Power

UPFC has the unique capability of independently controlling both real power flow (P) on a transmission line & the reactive power (Q) [6] at a specified point. The transmission line containing the UPFC thus appears to the rest of the power system as a high impedance power source or sink. This is an extremely powerful mode of operation that has not previously been achievable with conventional line compensating equipment.

3.3 Reliability Logic Diagram Using Series – Parallel System

The reliability logic diagram of Unified Power Flow Controller and Series Compensator using Series – Parallel system is shown in Fig. 3.2. Each rectangle block in the figure represents a particular component. Here, each component has its reliability which is independent of the time. Considering these reliabilities and in combination of simple series and parallel system, the overall reliability and unreliability of the system are determined as follows:

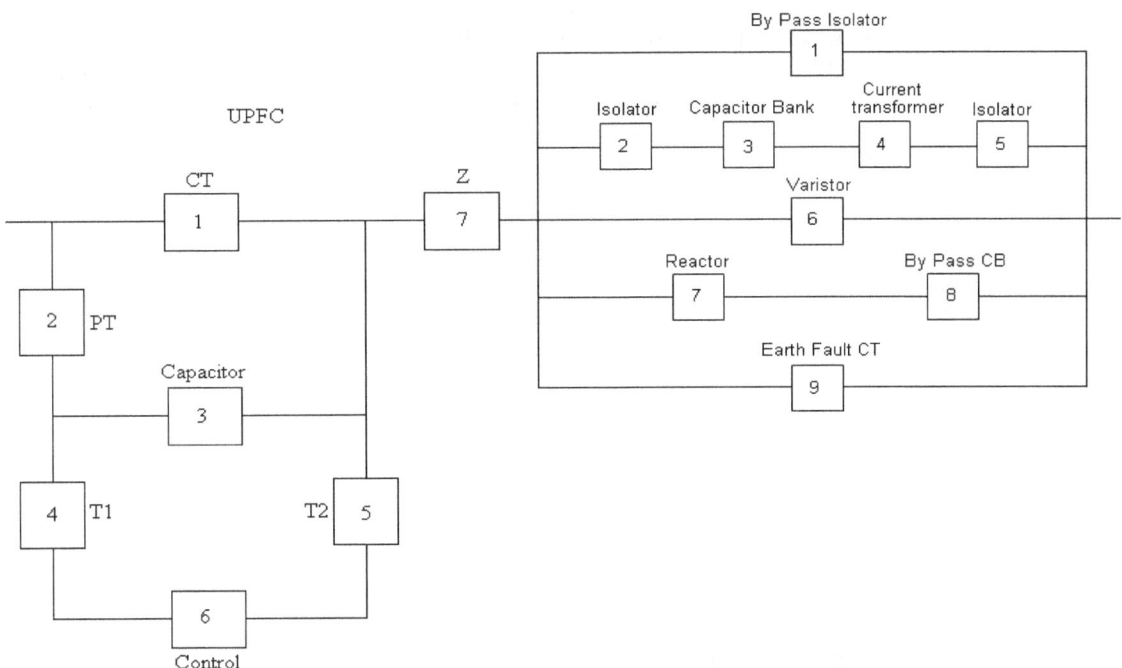

Fig. 3.2: RLD for combination of UPFC and Series Compensator using Series – Parallel System

For UPFC:
    Block 4, 5 & 6 are in series
$$R_{8`} = R_{4`} * R_{5`} * R_{6`}$$
    Block 3 & 8 is in parallel
$$R_{9`} = 1 - Q_{3`} * Q_{8`} \quad\quad Q_{3`} = 1 - R_{3`} \;\&\; Q_{8`} = 1 - R_{8`}$$
    Block 9 & 2 is in series
$$R_{10`} = R_{2`} * R_{9`}$$
    Block 1 & 10 is in Parallel
$$R_{11`} = 1 - Q_{1`} * Q_{10`} \quad\quad Q_{1`} = 1 - R_{1`} \;\&\; Q_{10`} = 1 - R_{10`}$$
    Block 11 & 7 is in series
$$R_{12`} = R_{7`} * R_{11`} \quad\quad\quad\quad\quad\quad\quad\quad\quad \ldots \quad (3.4)$$

For Series Compensator:
$$R_{10} = R_2 R_3 R_4 R_5$$
$$R_{11} = R_7 R_8$$
$$R_{12} = 1 - [(1 - R_1)(1 - R_{10})(1 - R_6)(1 - R_{11})(1 - R_9)]$$
$$Q_{10} = 1 - R_{10} \quad\quad Q_{11} = 1 - R_{11} \quad\quad Q_{12} = 1 - R_{12} \quad \ldots \quad (3.5)$$

For UPFC and Series Compensator:
$$R_{13} = R_{12}^1 R_{12} \quad\quad\quad \ldots \quad (3.6)$$

Where R_{12}^1 is the reliability of UPFC, R_{12} is the reliability of Series Compensator and R_{13} is the reliability for the combination of UPFC and Series Compensator. 'Q' represents the unreliability of the particular component or system.

3.3.1 Results

Considering individual reliabilities of each component as given in Section 2.2.1, and the remaining components data is as follows:

    Capacitor [UPFC] = 0.85   Thyristor = 0.78
    UPFC Control = 0.94   Potential Transformer = 0.92

Substituting all the reliability values in the Eqns. (3.4) to (3.6),
    Reliability of the configuration shown in Fig. 3.2 is = **0.9751824**
    Unreliability of the configuration shown in Fig. 3.2 is = **0.0248176**

3.4 Reliability Logic Diagram Using State Space representation

The State Space representation for the combination of UPFC and Series Compensator is shown in Fig. 3.3, where U means upstate of a component, D means down state of the component, λ and μ are failure and repair rate of each component respectively which is considered to be same throughout the analysis.

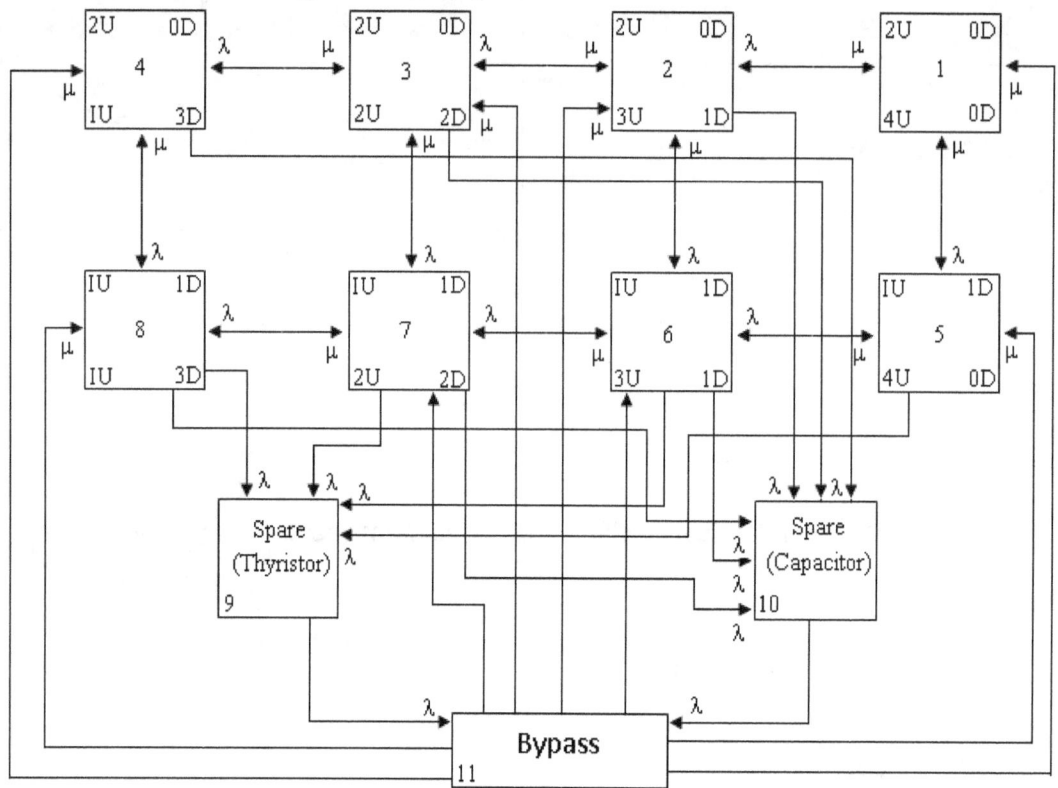

Fig. 3.3: RLD for combination of UPFC and Series Compensator using State – Space Representation

This is another method for finding the reliability of entire system [21]. Here the rectangular blocks from 1 to 8 represents the transition state with combination of UPFC and series compensator. The upper transition represents the states of UPFC and lower transitions represent states of Series Compensator. Here, only 8 states are considered for the combination of both elements and the remaining is not considered because, the remaining states cannot with stand the rated capacity of the transmission line.

The state space diagram consists of two spares one of thyristor and the other of Capacitor, because, each state is a combination of these two elements. Each and every state is connected to bypass module [9], because, at any state the system can be failed due to any faults or improper firing of thyristors or failure of capacitors in series compensator. Each state is assigned with a proper transition number in a sequential manner so that, each state follows the previous sates. A bypass block is also considered because, to reduce the capacitor voltage which is due to fault currents.

3.4.1 Illustrations

From the above reliability logic diagram, the Stochastic Transitional Probability Matrix (STPM) P is:

$$P = \begin{bmatrix} 1-2\lambda & \lambda & 0 & 0 & \lambda & 0 & 0 & 0 & 0 & 0 & 0 \\ \mu & 1-(\mu+2\lambda) & \lambda & 0 & 0 & \lambda & 0 & 0 & 0 & \lambda & 0 \\ 0 & \mu & 1-(\mu+3\lambda) & \lambda & 0 & 0 & \lambda & 0 & 0 & \lambda & 0 \\ 0 & 0 & \mu & 1-(\mu+2\lambda) & 0 & 0 & 0 & \lambda & 0 & \lambda & 0 \\ \mu & 0 & 0 & 0 & 1-(\mu+2\lambda) & \lambda & 0 & 0 & \lambda & 0 & 0 \\ 0 & \mu & 0 & 0 & \mu & 1-(2\mu+3\lambda) & \lambda & 0 & \lambda & \lambda & 0 \\ 0 & 0 & \mu & 0 & 0 & \mu & 1-(2\mu+3\lambda) & \lambda & \lambda & \lambda & 0 \\ 0 & 0 & 0 & \mu & 0 & 0 & \mu & 1-(2\mu+2\lambda) & \lambda & \lambda & 0 \\ 0 & 0 & 0 & 0 & 0 & 0 & 0 & 0 & 1-\lambda & 0 & \lambda \\ 0 & 0 & 0 & 0 & 0 & 0 & 0 & 0 & 0 & 1-\lambda & \lambda \\ \mu & \mu & \mu & \mu & \mu & \mu & \mu & \mu & 0 & 0 & 1-8\mu \end{bmatrix}$$

$$[P_{SS}][P] = [P_{SS}]$$

Where, $P_{SS} = [P_1\ P_2\ P_3\ \ldots\ P_9\ P_{10}\ P_{11}]$ which is a limiting state probability vector.

Expressing the above matrix form in terms of equations,

$$P_1(1-2\lambda) + P_2\mu + P_5\mu + P_{11}\mu = P_1 \quad \ldots \quad (3.7)$$

$$P_1\lambda + P_2(1-(\mu+2\lambda)) + P_3\mu + P_6\mu + P_{11}\mu = P_2 \quad \ldots \quad (3.8)$$

$$P_2\lambda + P_3(1-(\mu+3\lambda)) + P_4\mu + P_7\mu + P_{11}\mu = P_3 \quad \ldots \quad (3.9)$$

$$P_3\lambda + P_4(1-(\mu+\lambda)) + P_8\mu + P_{11}\mu = P_4 \quad \ldots \quad (3.10)$$

$$P_1\lambda + P_5(1-(\mu+2\lambda)) + P_6\mu + P_{11}\mu = P_5 \quad \ldots \quad (3.11)$$

$$P_2\lambda + P_5\lambda + P_6(1-(2\mu+3\lambda)) + P_7\mu + P_{11}\mu = P_6 \quad \ldots \quad (3.12)$$

$$P_3\lambda + P_6\lambda + P_7(1-(2\mu+3\lambda)) + P_8\mu + P_{11}\mu = P_7 \quad \ldots \quad (3.13)$$

$$P_4\lambda + P_7\lambda + P_8(1-(2\mu+2\lambda)) + P_{11}\mu = P_8 \quad \ldots \quad (3.14)$$

$$P_5\lambda + P_6\lambda + P_7\lambda + P_8\lambda + P_9(1-\lambda) = P_9 \quad \ldots \quad (3.15)$$

$$P_2\lambda + P_3\lambda + P_4\lambda + P_6\lambda + P_7\lambda + P_8\lambda + P_{10}(1-\lambda) = P_{10} \quad \ldots \quad (3.16)$$

$$P_9\lambda + P_{10}\lambda + P_{11}(1-8\mu) = P_{11} \quad \ldots \quad (3.17)$$

Since all the above Eqns. (3.7) to (3.17) are independent to each other, we consider only 10 equations out of the above 11 equations and 11th Eqn. is to be taken as

$$P_1+P_2+P_3+\ldots\ldots\ldots+P_9+P_{10}+P_{11} = 1 \quad \ldots \quad (3.18)$$

Writing the above Eqns (3.7) to (3.16) and (3.18) in matrix form,

$$\begin{bmatrix}
-2\lambda & \mu & 0 & 0 & \mu & 0 & 0 & 0 & 0 & 0 & \mu \\
\lambda & -(\mu+2\lambda) & \mu & 0 & 0 & \mu & 0 & 0 & 0 & 0 & \mu \\
0 & \lambda & -(\mu+3\lambda) & \mu & 0 & 0 & \mu & 0 & 0 & 0 & \mu \\
0 & 0 & \lambda & -(\mu+\lambda) & 0 & 0 & 0 & \mu & 0 & 0 & \mu \\
\lambda & 0 & 0 & 0 & -(\mu+2\lambda) & \mu & 0 & 0 & 0 & 0 & \mu \\
0 & \lambda & 0 & 0 & \lambda & -(2\mu+3\lambda) & \mu & 0 & 0 & 0 & \mu \\
0 & 0 & \lambda & 0 & 0 & \lambda & -(2\mu+3\lambda) & \mu & 0 & 0 & \mu \\
0 & 0 & 0 & \lambda & 0 & 0 & \lambda & -(2\mu+2\lambda) & 0 & 0 & \mu \\
0 & 0 & 0 & 0 & \lambda & \lambda & \lambda & \lambda & -\lambda & 0 & 0 \\
0 & \lambda & \lambda & \lambda & 0 & \lambda & \lambda & \lambda & 0 & -\lambda & 0 \\
1 & 1 & 1 & 1 & 1 & 1 & 1 & 1 & 1 & 1 & 1
\end{bmatrix}
\begin{bmatrix} P_1 \\ P_2 \\ P_3 \\ P_4 \\ P_5 \\ P_6 \\ P_7 \\ P_8 \\ P_9 \\ P_{10} \\ P_{11} \end{bmatrix}
=
\begin{bmatrix} 0 \\ 0 \\ 0 \\ 0 \\ 0 \\ 0 \\ 0 \\ 0 \\ 0 \\ 0 \\ 1 \end{bmatrix}$$

... (3.19)

Based on the above equations, the solution methodology to find the Limiting State Probabilities of the states shown in Fig. 3.3 is presented as algorithm in the next section.

3.4.2 Algorithm

1. Start
2. Enter State transitional Probability Matrix P which is a square matrix of size n.
3. Define Limiting State Probability Vector P_{ss} which is a row matrix.
4. Solve $P_{ss} * P = P_{ss}$
5. From Step 4, get 'n' Eqns. out of which (n-1) will be independent to each other.
6. From Step 5, consider any n-1 equations.
7. Consider $P_1 + P_2 + P_3 + \ldots + P_n = 1$ as n^{th} equation.
8. Write Eqns. Obtained from Steps 6 & 7 in Matrix form.
9. Solve Eqns. of step 8 using either Cramer's rule or Gauss Elimination method, compute Limiting State Probabilities $P_1, P_2 \ldots P_n$.
10. Availability will be equal to the sum of the LSPs of the states which leads to upstate.
11. Unavailability will be equal to the sum of the LSPs of the states which leads to downstate.
12. Stop

3.4.3 Results

From the matrix form of Eqn. (3.19) the limiting state probabilities are obtained by considering the data: Failure Rate (λ) = 0.7 f/yr
Repair Rate (μ) = 150 hrs of each component, then

Individual LSPs are:

P_1 = 0.5631 P_2 = 0.0046 P_3 = 0.00025 P_4 = 1.2 e - 4
P_5 = 2.3 e - 5 P_6 = 11.2 e - 6 P_7 = 23.2 e - 6 P_8 = 13.2 e - 7
P_9 = 0.20915 P_{10} = 0.20915 P_{11} = 12.8 e - 9

Therefore, the sum of the limiting state probabilities is
$P_1 + P_2 + P_3 + P_4 + P_5 + P_6 + P_7 + P_8 + P_9 + P_{10} + P_{11} = 1$

Availability of the system (P_{UP}) = $P_1 + P_9 + P_{10}$ = 0.5631 + 0.20915 + 0.20915 = **0.995**
Unavailability (P_{DOWN}) = 1 - 0.995 = **0.005**

3.5 Load Indices

The availability of transmission line and the associated transfer capability are modified [12] using the reliability models when the UPFC is in service. The failure rate and the repair time of the capacitor bank in each module are 0.7f/yr and 150hr respectively. Reliability analysis of the entire transmission system is being carried out by using load indices [11] like Loss of Load Expectation (LOLE) and Loss of Energy Expectation (LOEE) [25].

The variations in the LOEE and LOLE [22] with system peak load are shown in Fig. 3.4 and 3.5 respectively. The annual load factor is assumed to be 70%. It can be seen that, for a give peak load, the LOEE and LOLE decrease with the employment of the UPFC. The inclusion of the UPFC allows the capacity of transmission line to be extended to its thermal limit [7] and therefore, transfer more available capacity from the remote generation site to the load point. The inclusion of series compensator allows reducing the effect of inductance.

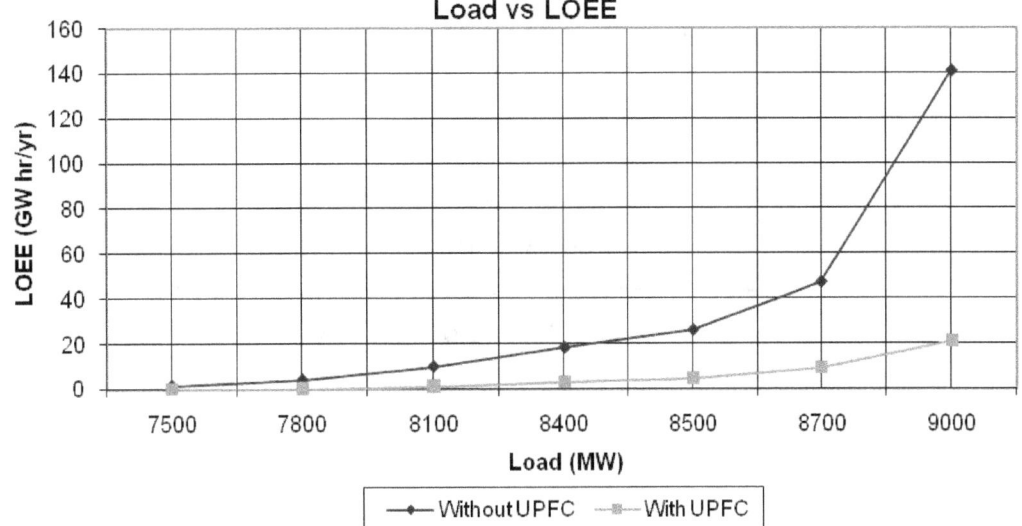

Fig. 3.4: Peak Load (MW) vs LOEE (GWhr/yr) wrt UPFC

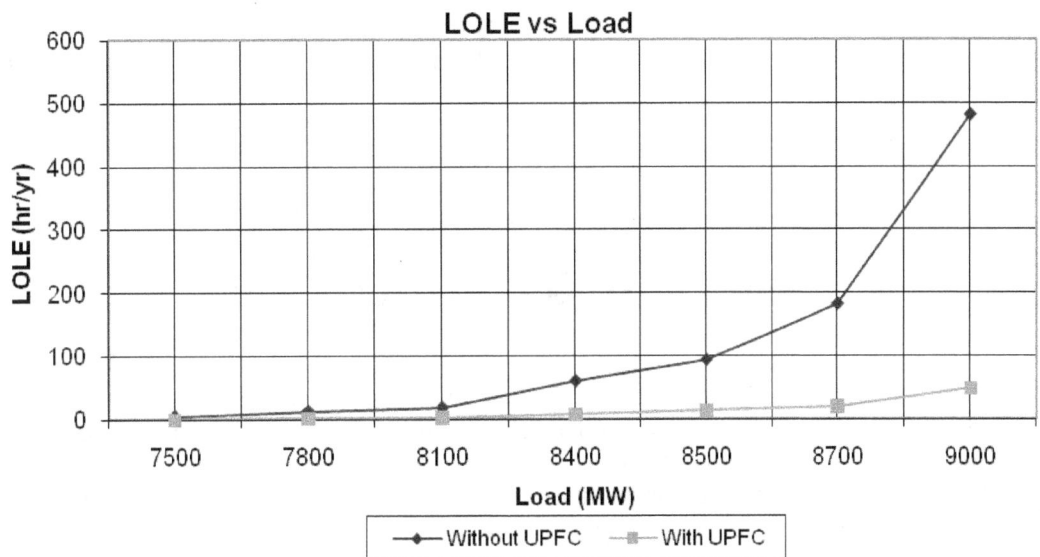

Fig. 3.5: Peak Load (MW) vs LOLE (hr/yr) wrt UPFC

Addition of capacitance in series with the transmission line modifies the reactance of the line. The difference is very small for lower system peak loads, as the local generation has sufficient capacity to prevent load interruption. The UPFC effect becomes significant as the peak load increases.

The impact of TCSC and series compensator on the LOEE and LOLE are shown in Figs. 3.6 and 3.7 respectively. The system load factor is varied from 50% to 100% in steps of 10%. It can be seen from these Figs. 3.6 & 3.7 that the LOEE and LOLE increases as the system load factor increases.

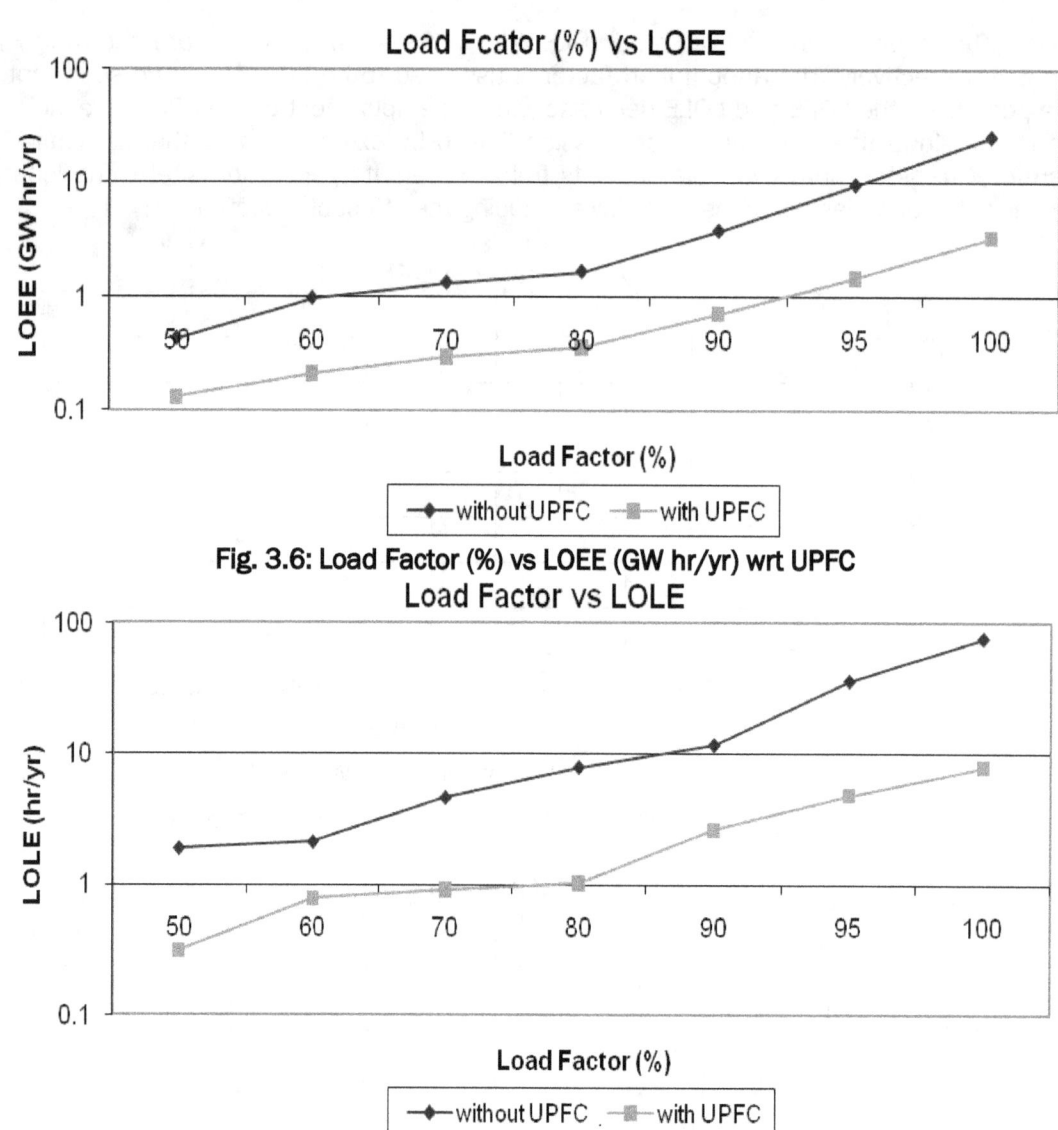

Fig. 3.6: Load Factor (%) vs LOEE (GW hr/yr) wrt UPFC

Fig. 3.7: Load Factor (%) vs LOLE (hr/yr) wrt UPFC

Comparison has been carried out between with and without UPFC and TCSC along with series compensator which is presented in Tables 3.1 to 3.4 respectively. The variation of Loss of Load Expectation with the system Load is shown in Table 3.1. Variation of Loss of Energy Expectation with the system load in discussed in Table 3.2. The variation of Loss of Load Expectation with the Load Factor is discussed in Table 3.3. Variation of Loss of Energy Expectation with the Load Factor is shown in Table 3.4.

Table 3.1: Load (MW) vs LOLE (hr/yr) wrt UPFC

LOLE			
Load MW	without UPFC	with UPFC	UPFC & SC
7800	13	1.4	0.3
8100	18	2.6	0.7
8400	60	7.2	1.5
8500	93	12.6	2.8
8700	182	20.1	6.5
9000	481	48	13.2

Table 3.2: Load (MW) vs LOEE (GW hr/yr) wrt UPFC

LOEE			
Load MW	without UPFC	with UPFC	UPFC & SC
7500	1.6	0	0
7800	4.2	0.2	0
8100	9.92	1.2	0.15
8400	18.3	3.1	0.5
8500	25.97	4.7	0.9
8700	47.26	9.12	1.4
9000	141.13	21.16	3.8

Table 3.3: Load Factor vs LOLE (hr/yr) wrt UPFC

LOLE			
Load Factor %	Without UPFC	with UPFC	UPFC & SC
50	1.89	0.312	0.001
60	2.11	0.78	0.12
70	4.62	0.91	0.24
80	7.917	1.025	0.4
90	11.67	2.63	0.68
95	35.8	4.81	0.94

Table 3.4: Load Factor vs LOEE (GW hr/yr) wrt UPFC

Load Factor %	LOEE		
	without UPFC	with UPFC	UPFC & SC
50	0.427	0.13	0.012
60	0.963	0.21	0.1
70	1.32	0.296	0.21
80	1.654	0.36	0.28
90	3.81	0.71	0.54
95	9.62	1.44	1.1

From Tables 3.1 and 3.2, it can be observed that, for any load LOLE & LOEE is decreasing for different configurations viz. without UPFC, with UPFC and UPFC & Series Compensator (SC), which indicates that the combination UPFC & SC is preferable for reduction of losses either in load or in energy indices. Similarly from Tables 3.3 and 3.4, instead of load (MW) load factor (%) of the system is considered and the reduction in losses can be observed. The graphical form of Tables 3.1 to 3.4 is represented in Figs. 3.4 to 3.7 respectively.

3.6 Comparison of Load Indices

Comparison has been carried out for both TCSC & UPFC along with series compensator. It was found that transmission system connected in series with UPFC has a very less loss in load & energy indices which can be seen from the Figs. 3.8 to 3.11 respectively.

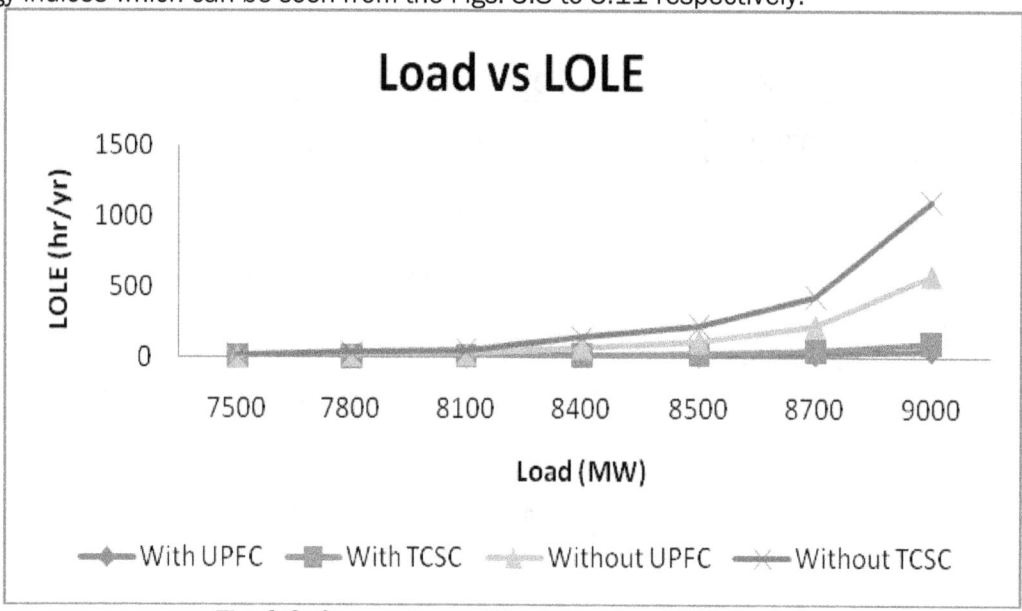

Fig. 3.8: Comparison of Peak Load vs LOLE (hr/yr)

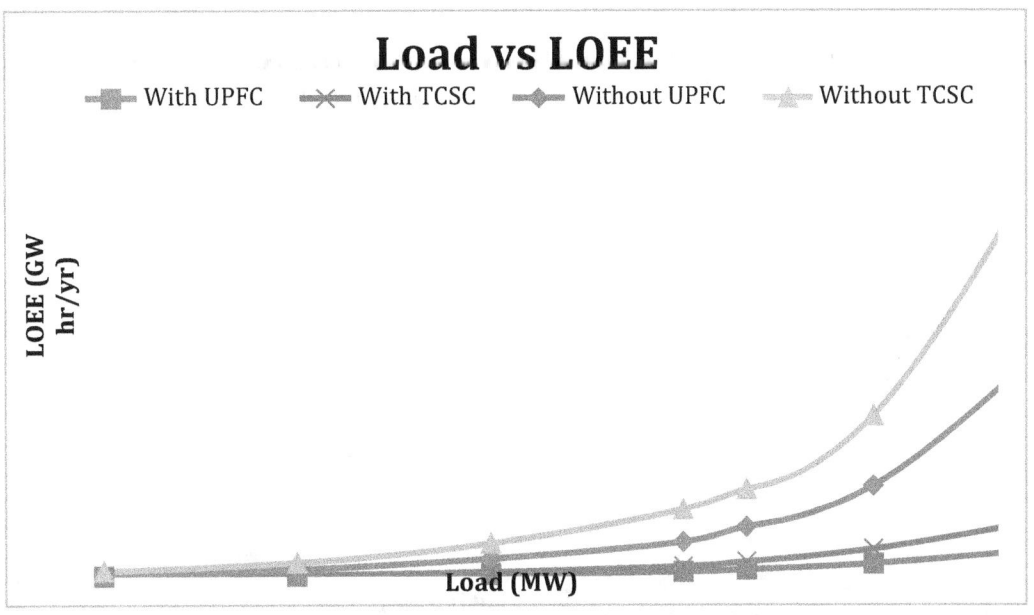

Fig. 3.9: Comparison of Peak Load vs LOEE (GW hr/yr)

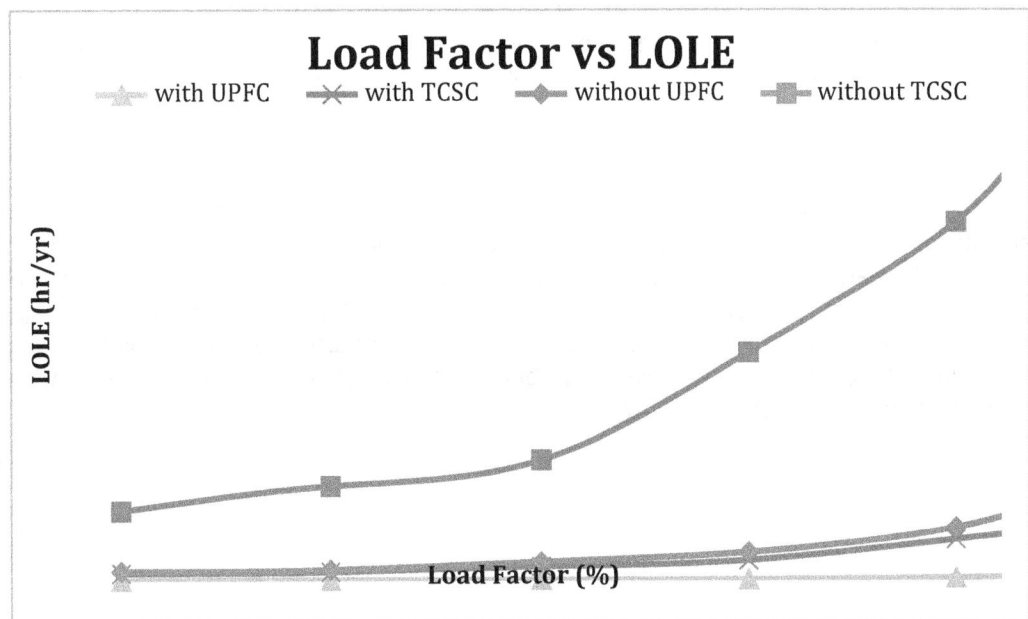

Fig. 3.10: Comparison of Load Factor vs LOLE (hr/yr)

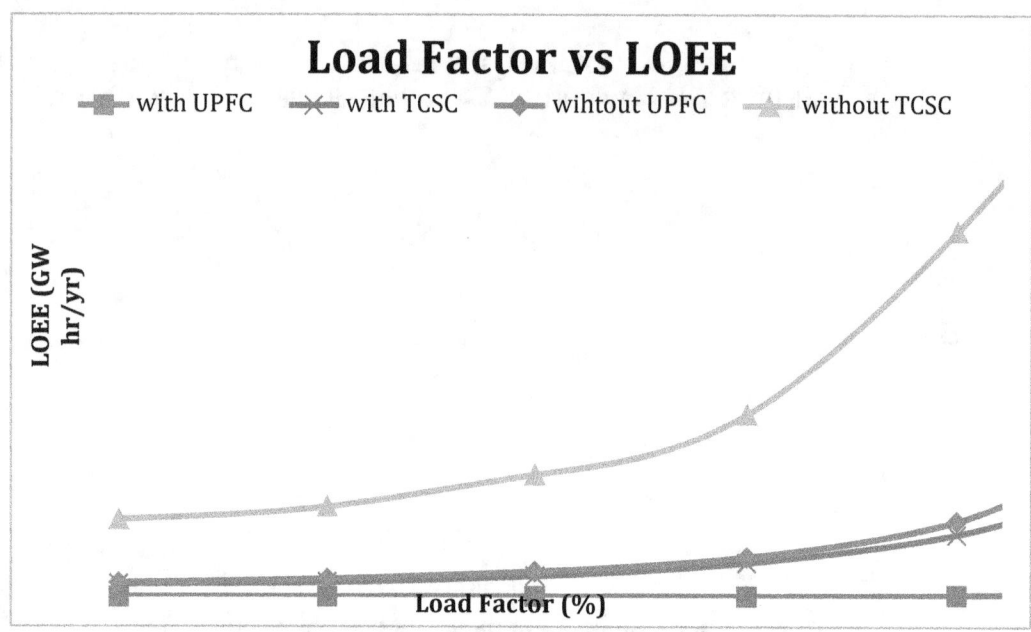

Fig. 3.11: Comparison of Load Factor vs LOEE (GW hr/yr)

From Figs. 3.8 to 3.11 it can be observed that LOLE & LOEE with respect to Load Factor and Load is decreasing when UPFC is incorporated into the system rather than incorporating TCSC.

3.7 Conclusions

In this chapter, reliability analysis of UPFC and series compensator is estimated with deterministic probability values with the available component data. Further, an attempt also has been made to estimate the availability of the combination of UPFC and series compensator using time dependent probabilities. These time dependent probabilities assume exponential distribution of the failure of the components whereas the deterministic probability data is based on binomial distribution. A comparison has been carried out between TCSC, UPFC when using series compensator between load indices.

Chapter 4

RELIABILITY ANALYSIS OF COMPOSITE POWER SYSTEM USING UPFC

4.1 Introduction

A UPFC consists of two switching converters where each converter is a voltage source inverter using gate turn off thyristor valves as shown in Fig. 4.1. UPFC functions as an ideal AC to AC power converter in which the real power can flow freely in either direction between the AC terminals of the two inverters and each inverter can independently generate reactive power at its own AC output terminals. Inverter 2 provides the main function of the UPFC by injecting an AC voltage V_1 with controllable magnitude & phase angle α at the power frequency, in series with the line can be considered essentially as a synchronous [29] AC voltage source. The real power exchanged at the AC terminal is converted by the inverter into DC power which appears at the DC link as positive & negative real power demand.

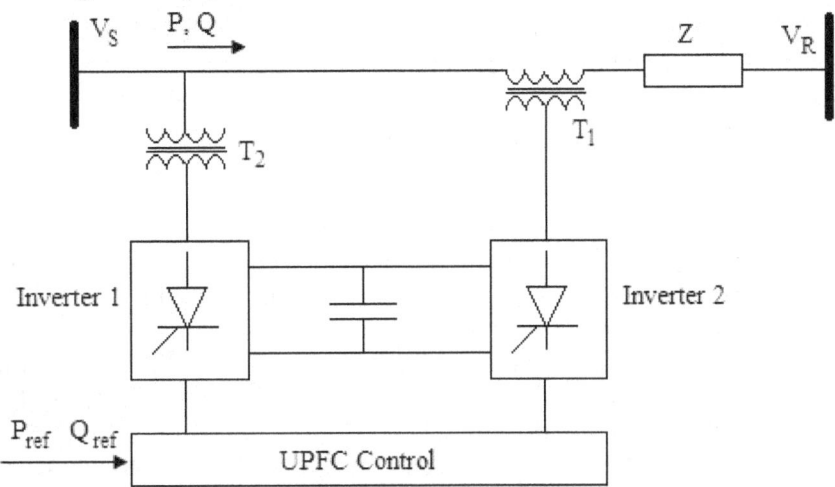

Fig. 4.1: Basic UPFC Model

The transmitted power P and the reactive power Q, supplied by the receiving end, can be expressed [9] as follows:

$$P - jQ = V_r \left(\frac{V_s + V_{pq} - V_r}{jX} \right)^* \qquad \qquad (4.1)$$

Where P = Active Power, Q = Reactive Power
V_r = Receiving end terminal voltage
V_s = Sending end terminal voltage
V_p = Injected voltage
X = Reactance of the line

The transmittable real power P is

$$P_O(\delta) - \frac{VV_{pq\,max}}{X} \le P_O(\delta) \le P_O(\delta) + \frac{VV_{pq\,max}}{X} \qquad \qquad (4.2)$$

Reactive Power Q is

$$Q_O(\delta) - \frac{VV_{pq\,max}}{X} \le Q_O(\delta) \le Q_O(\delta) + \frac{VV_{pq\,max}}{X} \qquad \qquad (4.3)$$

where δ = Transmission Angle
$P_O(\delta)$ = Normalized transmitted Power
$Q_O(\delta)$ = Normalized reactive Power

UPFC has the unique capability of independently controlling both real power (P) flow on a transmission line & the reactive power (Q) [26] at a specified point. The transmission line containing the UPFC thus appears to the rest of the power system as a high impedance power source or sink. This is an extremely powerful mode of operation that has not previously been achievable with conventional [31] line compensating equipment.

Composite Power System constitutes of both Generation and Transmission facilities including their auxiliary components.

In this Chapter, the reliability analysis of composite power system using UPFC is presented. The reliability analysis viz. Availability and Unavailability for multi-module UPFC is discussed by using State Space representation in Section 4.2. The reliability analysis viz. System Indices (BPSD, BPII and BPECI), Probability of Failure and EENS of IEEE 6 Bus Roy Billinton Test System (RBTS) is discussed in Section 4.3 with all results. Conclusions of Chapter 4 are presented in Section 4.4.

4.2 Reliability Analysis of UPFC

The reliability evaluation of a composite power system involves four key steps:
1. Reliability modeling of the generation & transmission units
2. Enumeration of all possible system contingencies
3. Determination of load curtailment under each contingency and
4. Calculation of the reliability indices at each load point.

First & third steps have been extended in order to incorporate FACTS in the overall evaluation.

The contingencies that are considered in point 2 are Under Line Outage Contingency. With the increased loading of existing power transmission system and the inclination towards maximizing economic benefits has led power system utilizes to run close to the limits of stable operation. The increase in power flows and ecological constraints have forced electric utilities to install new equipment to enhance network process. If a UPFC comes into operation with appropriate control parameters, it is possible to improve the line flow capacity of few such lines and thereby avoiding the possible voltage collapse. Steady state security assessment enables the operating personnel to know which system disturbances or contingencies may cause limit violations and force the system to enter into emergency state.

The reliability of the composite power system is determined by System [16] & Load indices which mainly concentrated on the increased capacity of UPFC [43] in the literature available so far. In this thesis, it is mainly concentrated on the state space representation of the test system which has not been carried out in the literature so far.

In reliability studies, the UPFC itself is represented by a two state model. In the UP state, the UPFC is capable of providing load flow control [38] & maximum transmission capacity. In the DOWN state, the UPFC is bypassed by a fully reliable circuit breaker and the system operates as a normal transmission line as shown in Fig. 4.2.

The probability and frequency value associated with each state are derived, independently and is presented here:

$$P(up) = \frac{A}{D} \qquad \qquad \ldots \qquad (4.4)$$

$$P(derated) = \frac{B}{D} \qquad \qquad \ldots \qquad (4.5)$$

$$P(down) = \frac{C}{D} \qquad \qquad \ldots \qquad (4.6)$$

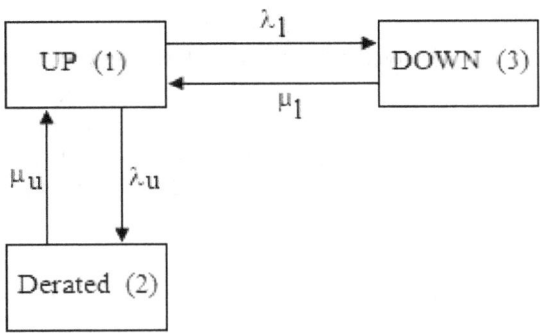

Fig. 4.2: State Space Model of the transmission element with derated state

$$F(up) = P(up) * (\lambda_1 + \lambda_u) \qquad \ldots \quad (4.7)$$
$$F(derated) = P(derated) * \mu_u \qquad \ldots \quad (4.8)$$
$$F(down) = P(down) * \mu_1 \qquad \ldots \quad (4.9)$$

where,
$$A = \mu_u \mu_1$$
$$B = \lambda_u \mu_1$$
$$C = \lambda_1 \mu_u$$
$$D = A + B + C$$

where, $\lambda_1, \mu_1, \lambda_u$ & μ_u are the failure and repair rate of the transmission line and UPFC respectively.

In Fig. 4.3, the state space representation of single module UPFC is shown. Each state in Fig. 4.3 is represented by rectangle block which enclosed left side number which is state number. Emergency state [18], state 3 is a transition state between states 1, 2 & 4. The number of states is increased as the number of modules increases.

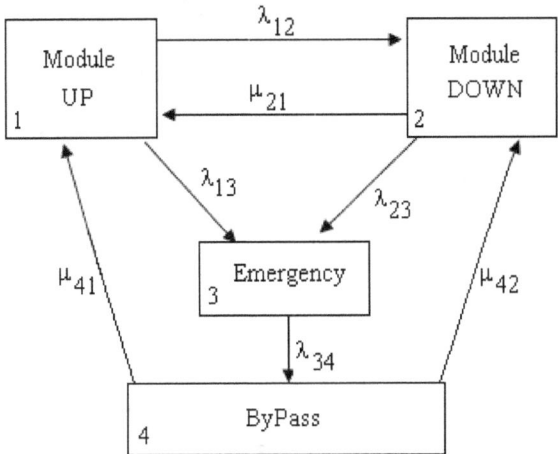

Fig. 4.3: State Space Representation of UPFC

Fig. 4.3 represents the state space representation of single module UPFC. A basic Single Module UPFC is shown in Fig. 4.1. Each state is represented by rectangle block enclosed with a number on left side which is defined as state number. A control circuit is present in the UPFC model so that all the components are controlled by using the controller. If the controller fails the UPFC goes into bypass state indicated as 4. The entire system may be in state 3 when the bypass (i.e. tripping circuits, relays etc.) fails, where the component is replaced by using a spare component. If the entire module is failed then the system will be in State 2 which is down state of the module indicated as 2.

A control circuit is present in the UPFC model so that all the components are controlled by using the controller. If the controller fails the UPFC goes into bypass [14] state. The entire system may be in state 3 when the bypass (i.e. tripping circuits, relays etc.) fails, where the component is replaced by using a spare component. As one module is insufficient for power transmission, analysis is carried out for different modules like 2, 3 etc.

Modules in this chapter indicate are One Single UPFC. The Basic Model of UPFC shown in Fig. 4.1 comprises of One Module. For a Two Module, Two basic model of UPFC are put together in series combination. Similarly, three modules and four modules etc.

The state space diagram of 2 modules UPFC with bypass and emergency states is shown in Fig. 4.4.

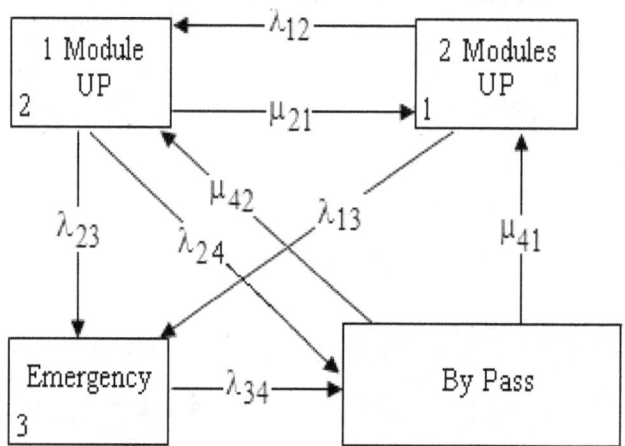

Fig. 4.4: State space diagram of 2 modules UPFC

The state space diagram of 3 modules UPFC with bypass and emergency states is shown in Fig. 4.5.

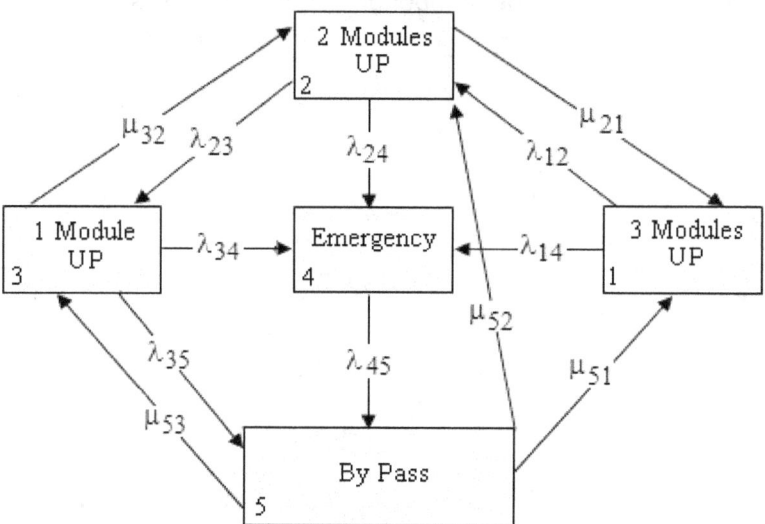

Fig. 4.5: State space diagram of 3 modules UPFC

The state space diagram of 4 modules UPFC with bypass and emergency states is shown in Fig. 4.6.

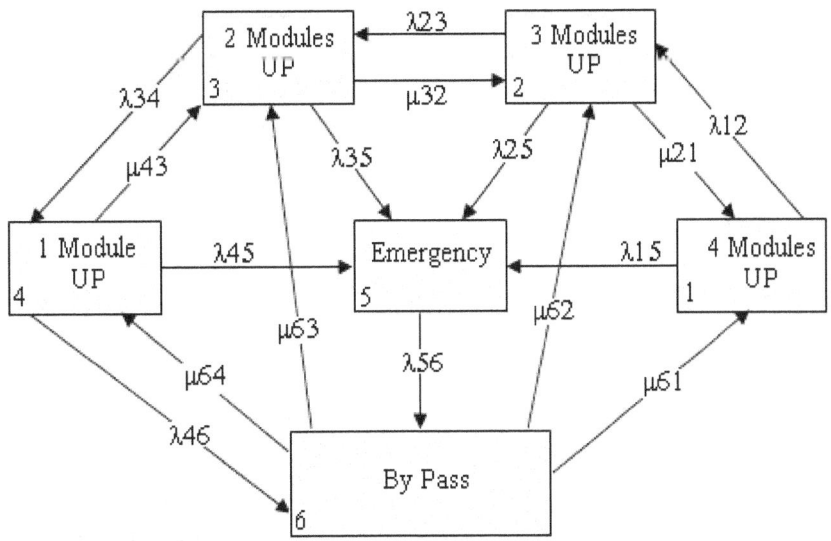

Fig. 4.6: State space diagram of 4 modules UPFC

The state space diagram of 5 modules UPFC with bypass and emergency states is shown in Fig. 4.7.

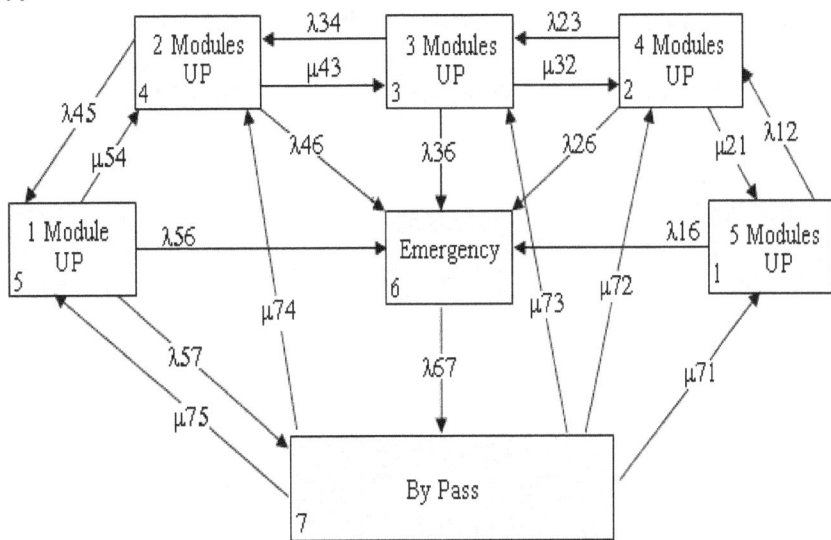

Fig. 4.7: State space diagram of 5 modules UPFC

The state space representation diagram of 6 modules UPFC with bypass and emergency states is shown in Fig. 4.8.

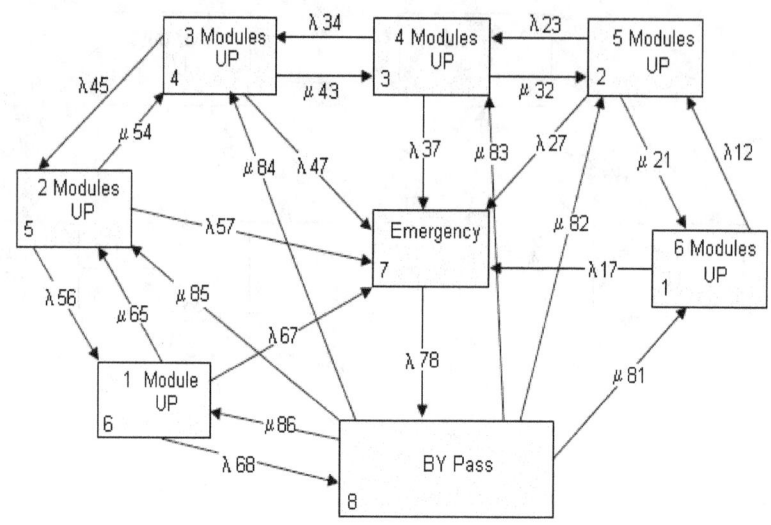

Fig. 4.8: State space diagram of 6 modules UPFC

The state space representation of seven modules UPFC with emergency & bypass is shown in Fig. 4.9.

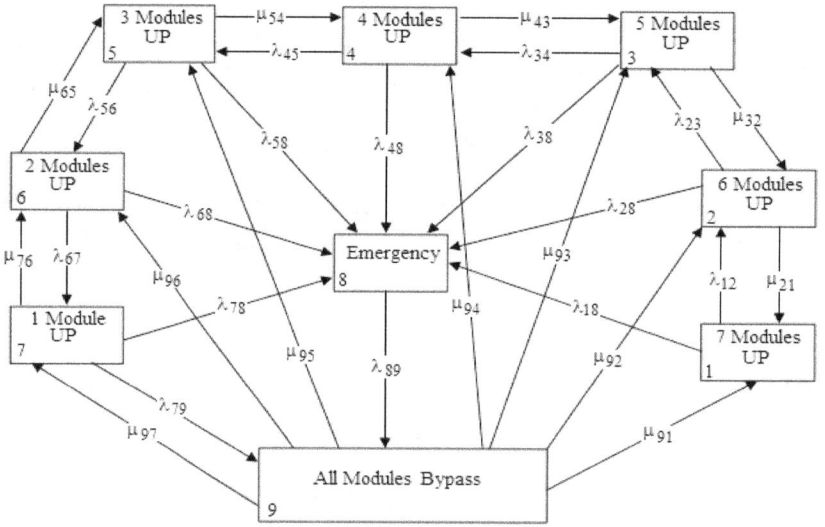

Fig. 4.9: State space representation of 7 Modules UPFC

The availability and Unavailability of UPFC by considering different modules from 2 to 8 have been obtained and presented in Table 4.1. From the results, as the number of modules increases the availability decreases.

Table 4.1: Availability & Unavailability for different Modules of UPFC

No. of Modules	Availability	Unavailability
2	0.99818	0.00182
3	0.99623	0.00377
4	0.99462	0.00538
5	0.99374	0.00626
6	0.99246	0.00754
7	0.99164	0.00836
8	0.98254	0.01746

From Table 4.1, it can be observed that as number of modules increases the availability of the system decreases.

4.3 Roy Billinton Test System

In the reliability analysis of a Composite Power system a 6 bus RBTS is considered as shown in Fig. 4.10

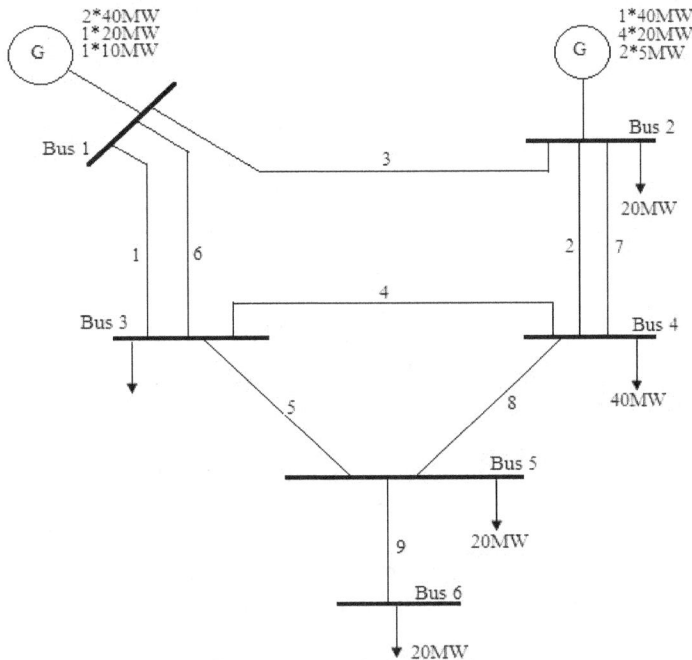

Fig. 4.10: Single Line diagram of 6 Bus RBTS

6 Bus RBTS is having 11 Generating units with 9 transmission lines & 5 Load points with 6 buses. The detailed information of RBTS is shown in Appendix - I.

4.3.1 Probability & Frequency:

A number of indices have been introduced in reliability theory to facilitate reliability predictions [49], and others to fit various applications. Quite generally, all of these indices can be classified into the following categories, such as *Probabilities, Frequencies, Mean Durations and Expectations*.

Some of these indices, the frequencies for example, apply to repairable components [53] and systems only. The expectations may be difficult to calculate and are, therefore, not as often used as the others; however, they include indices, such as the first above, that are able to 'weigh' failures by taking on higher values for failures with serious consequences than for failures that are marginal.

The probability of each of state of system can be found out by using the relation [27]

$$^nC_r A^{n-r} U^r. \qquad (4.10)$$

where A = Availability of the System
U = Unavailability of the system
n = No. of generators of the system
r = States of No. of generators on Outage

Similarly the frequency of the system can be calculated as:

Frequency (occ / year) = Probability * departure rate (occ / year).   . .   (4.11)

In Table 4.2, the probability & frequency of occurrence / year of a 6 Bus RBTS when considering no. of outages in generation are presented, with the data assumed is A = 0.98, U = 0.02, n = 11, r = 0 to 11.

Table 4.2: Probability & Frequency of 6 BUS RBTS Generation

State	No. of Generators on Outage	Capacity Available	Probability	Dep. Rate (occ/year)	Frequency (occ/year)
1	0	240	0.80073135	20	16.014627
2	1	200	0.1797	180	32.346
3	2	160	0.01834	360	6.6024
4	3	120	0.001123	370	0.41551
5	4	100	0.0000458	500	0.0229
6	5	80	0.00000131	540	0.00070686
7	6	60	2.67×10^{-8}	700	1.869×10^{-5}
8	7	40	3.89×10^{-10}	760	2.9564×10^{-7}
9	8	20	3.975×10^{-12}	900	3.5775×10^{-9}
10	9	10	2.704×10^{-14}	1020	2.758×10^{-11}
11	10	5	1.1038×10^{-16}	1674	1.8477×10^{-13}
12	11	0	2.048×10^{-19}	1896	3.88×10^{-16}

The 6 Bus RBTS consists 9 transmission lines. The availability & unavailability of the transmission lines are calculated and presented in Table 4.3.

Table 4.3: Availability & Unavailability of 6 BUS RBTS Transmission Lines

Transmission Line	Availability	Unavailability
1	0.99636033	0.00363967
2	0.99545455	0.00454545
3	0.99658703	0.00341297
4	0.9943267	0.0056733
5	0.9921684	0.0078316
6	0.9958326	0.0041674
7	0.9973467	0.0026533
8	0.9936972	0.0063028
9	0.9924871	0.0075129

4.3.2 System Indices

In a more practical network there are a number of load points and each point has a distinct set of reliability indices. The basic parameters are the probability & frequency of failure at the individual load points, but additional indices can be created from these generic values. The individual load point indices can also be aggregated to produce system indices which include, in addition to consideration of generation adequacy, recognition of the need to move the generated energy through the transmission network to the customer load points.

If these indices are calculated for a single load level and expressed on a base of one year, they should be designated as *annualized* values. Annualized indices calculated at the system peak load level are usually much higher than the actual annual indices.

$$\text{BPSD} = \frac{\sum_k \sum_{j \in x,y} L_{kj} F_j}{\sum_{j \in x,y} F_j} \qquad \ldots \quad (4.12)$$

$$\text{Bulk Power Interruption Index (BPII)} = \frac{\sum_k \sum_{j \in x,y} L_{kj} F_j}{L_s} \qquad \ldots \quad (4.13)$$

$$\text{BPECI} = \frac{\sum_K \sum_{j \in x,y} L_{Kj} D_{Kj} F_j * 60}{L_s} \qquad \ldots \quad (4.14)$$

The system indices are calculated by using the above relations for a 6 Bus RBTS and presented in Table 4.4.

Table 4.4: System Indices for 6 Bus RBTS with original scheme

System Indices	RBTS
BPSD	19.71
BPII	0.367
BPECI	331.16

$$\text{BPSD} = \frac{88.08}{4.468} = 19.71 \text{ MW/disturbance}$$

$$\text{BPII} = \frac{88.08}{240} = 0.367 \text{ MW / MW-yr}$$

$$\text{BPECI (Severity Index)} = 60 * \frac{68.36 * 0.962192 * 20.14}{240} = 331.16 \text{ MWh/MW-yr}$$

The system indices for a 6 Bus RBTS are once again calculated by changing the number of modules of UPFC and are tabulated in Table 4.5 and represented in Figs. 4.11, 4.12 & 4.13 respectively, of the graphs BPSD, BPII and BPECI vs number of modules.

Table 4.5: System Indices of 6 Bus RBTS with different modules of UPFC

Module No	BPSD	BPII	BPECI
2	19.69	0.3642	331.14
3	19.58	0.363	331.11
4	19.47	0.3622	331.08
5	19.39	0.3612	331.04
6	19.31	0.361	330.91
7	19.26	0.359	330.69
8	19.38	0.364	331.01

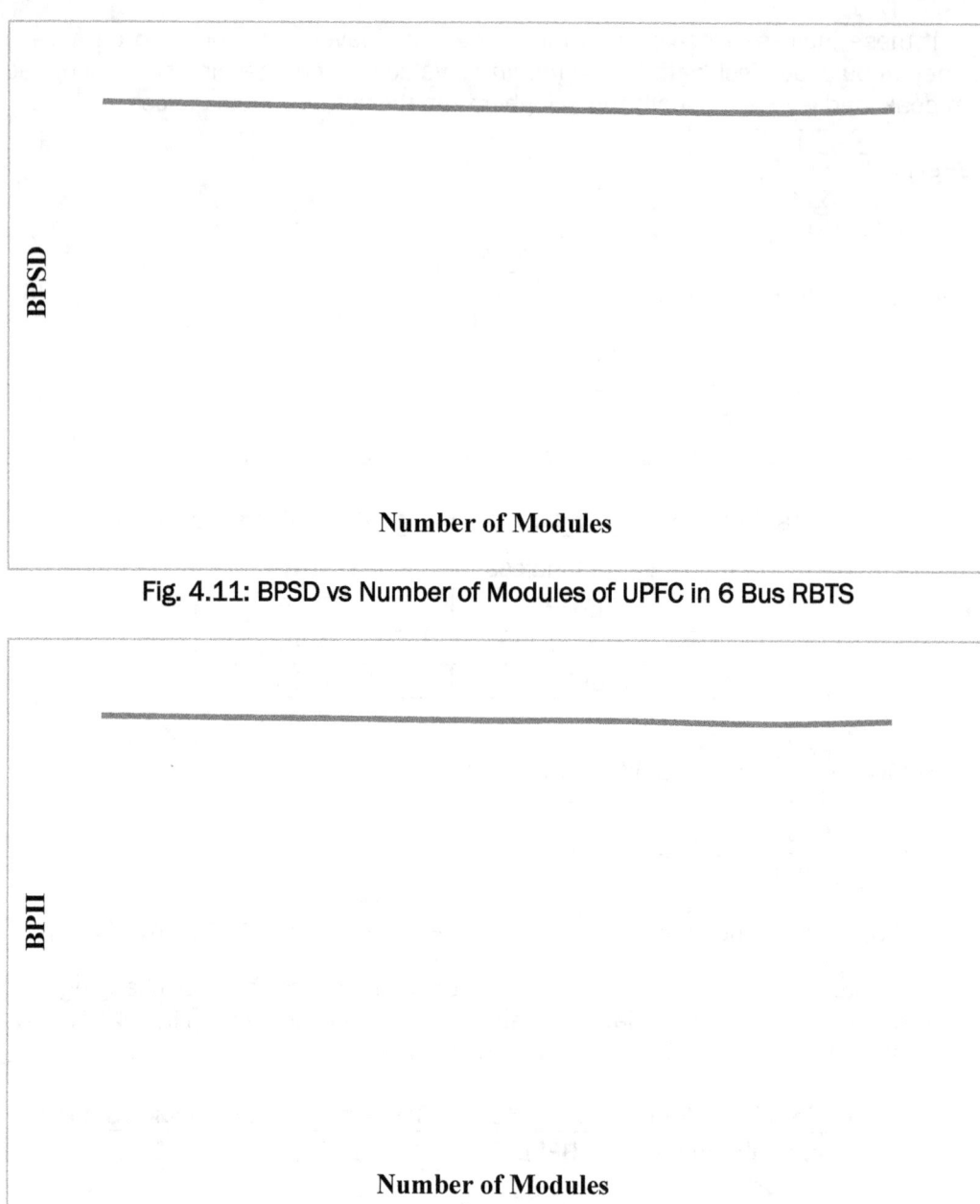

Fig. 4.11: BPSD vs Number of Modules of UPFC in 6 Bus RBTS

Fig. 4.12: BPII vs Number of Modules of UPFC in 6 Bus RBTS

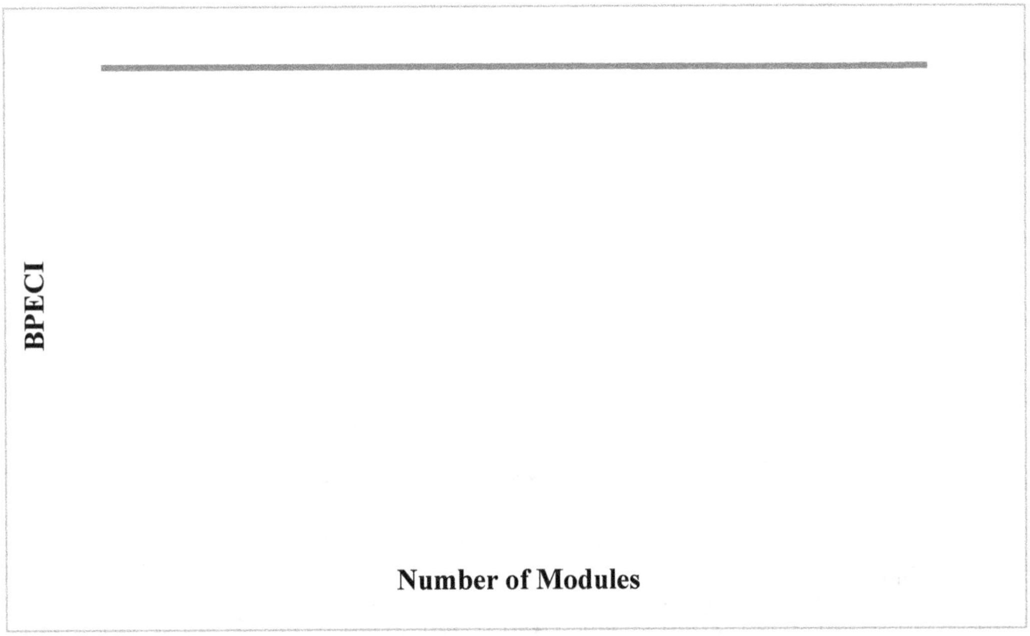

Fig. 4.13: BPECI vs Number of Modules of UPFC in 6 Bus RBTS

From Table 4.5, it can be observed that the system indices BPSD, BPII and BPECI are decreasing as the number of modules increasing. For the system considered the number of modules required is Seven. As the number of modules increases beyond the required number the system indices is also increasing, this is due to over thermal loading of FACTS devices beyond their limit, which can be observed from Table 4.5. The graphical representation of Table 4.5 can be observed in Figs. 4.11 to 4.13.

Systems indices BPSD, BPII & BPECI are calculated with respect to the generation capacity (MW) without considering any FACTS devices (UPFC) for 6 Bus RBTS. The system indices are tabulated in Table 4.6 and represented in Fig. 4.14. BPECI is also referred by some authors as Severity Index [35].

Table 4.6: System Indices of 6 Bus RBTS vs Generation Capacity – without UPFC

Generation Capacity (MW)	Load Demand (MW)	BPSD	BPII	BPECI
240	185	19.834	0.364	329.642
270	203.5	19.897	0.372	331.246
300	222	20.653	0.379	333.201
330	240.5	21.842	0.389	336.721
345	259	34.197	0.465	371.264
360	277.5	78.673	0.913	615.259

From Table 4.6, it can be observed that, system indices BPSD, BPII & BPECI increases as the Generation capacity and Load Demand of the system increases, without using any FACTS controlling devices in the given system. This is because, load at the given bus or transmission line can't be predicted directly. The graphical form of Table 4.6 is shown in Fig. 4.14.

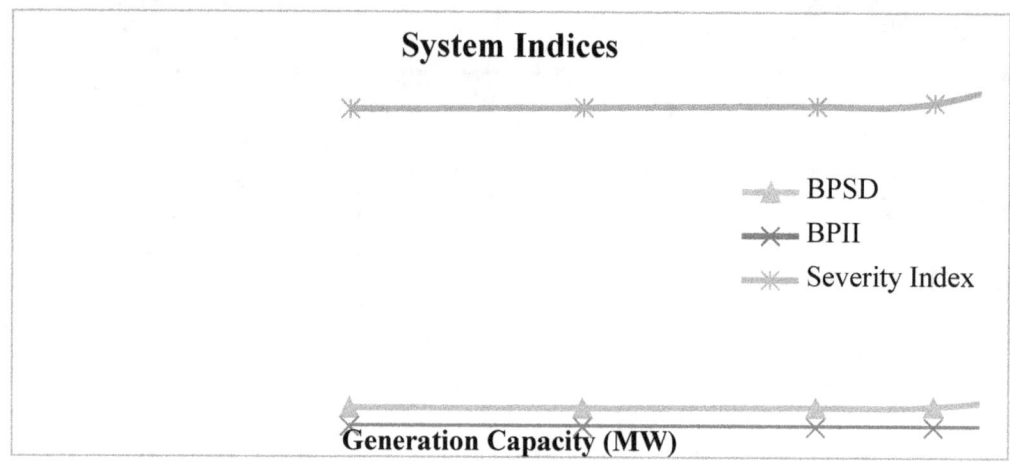

Fig. 4.14: System Indices vs Generation Capacity of 6 Bus RBTS

Systems indices via, BPSD, BPII & BPECI [58] are calculated with respect to the generation capacity (MW) by considering 7 modules UPFC for 6 Bus RBTS. The system indices are tabulated in Table 4.7. A comparison representation when using 7 module UPFC & without using UPFC are presented in Figs. 4.15, 4.16 & 4.17 respectively for these three indices.

Table 4.7: System Indices vs Generation Capacity – with 7 modules UPFC

Generation Capacity (MW)	Load Demand (MW)	BPSD	BPII	BPECI
240	185	19.428	0.324	328.521
270	203.5	19.679	0.334	330.125
300	222	20.315	0.351	331.921
330	240.5	21.242	0.360	332.671
345	259	23.157	0.378	333.164
360	277.5	29.736	0.389	334.259

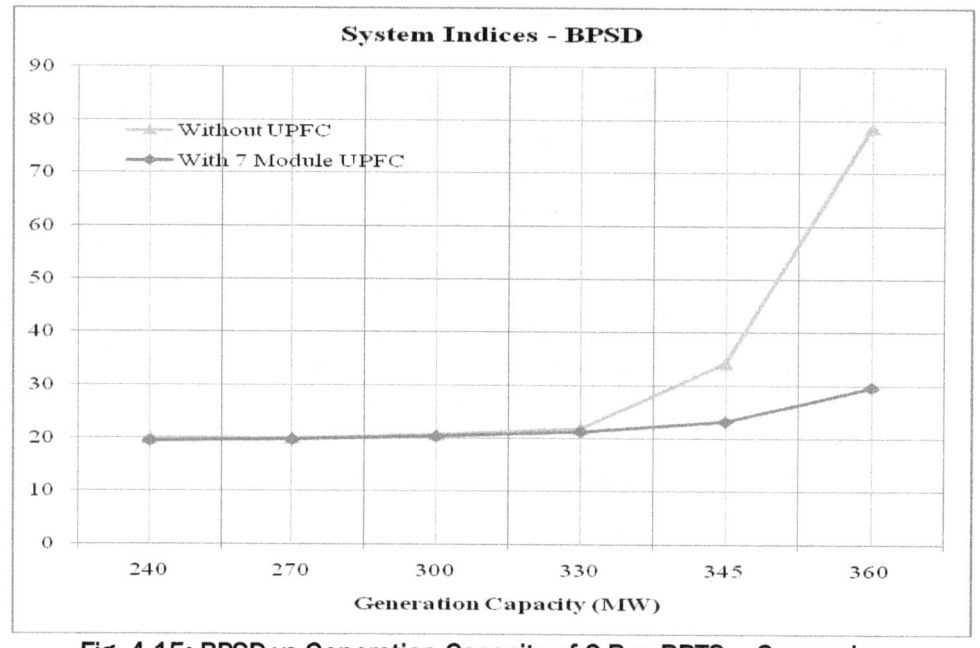

Fig. 4.15: BPSD vs Generation Capacity of 6 Bus RBTS – Comparison

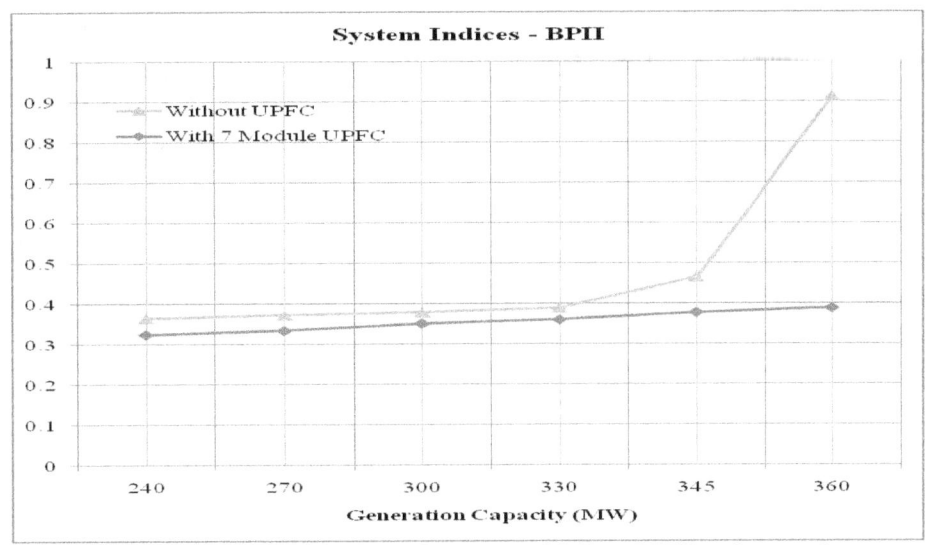

Fig. 4.16: BPII vs Generation Capacity of 6 Bus RBTS – Comparison

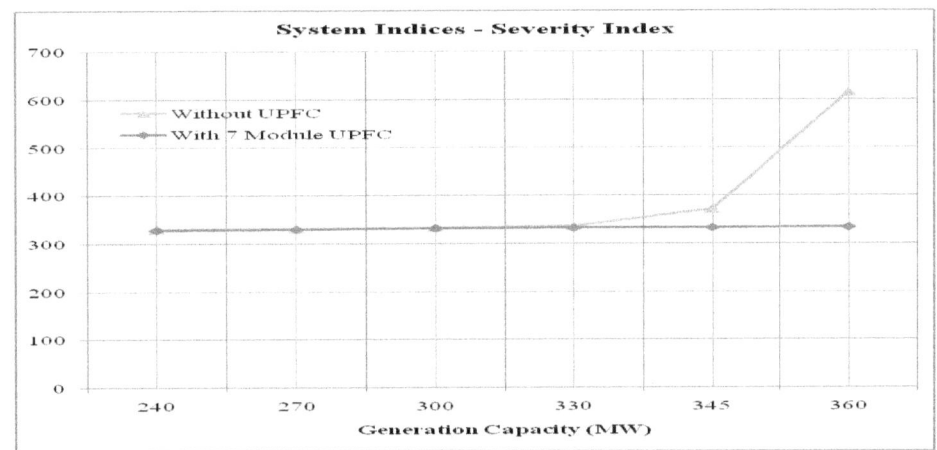

Fig. 4.17: Severity Index vs Generation Capacity of 6 Bus RBTS – Comparison

From Table 4.7, it can be noted that the system indices BPSD, BPII and BPECI are increasing as the generation capacity and load demand increases when Seven Module UPFC is considered. The graphical representation of Table 4.7 can be observed in Figs. 4.15 to 4.17 respectively.

From Fig. 4.15, 4.16 & 4.17 it can be seen that as the generating capacity (MW) is increasing the system indices like BPSD, BPII, BPECI are increasing rapidly after the desired generating capacity i.e. 330MW. When the system is not using any FCATS devices the supply disturbance, interruption in the supply & Severity Index are also increasing rapidly resulting in more losses and less efficiency of the system. When UPFC is incorporated into the system with 7 modules the system indices are gradually decreased resulting in higher efficiency.

Whenever the number of components / elements increases obviously the reliability of the system decrease as discussed earlier. In order, to improve the reliability of the system it is proposed to increase the capacity of the UPFC so that the number of modules can be reduced. An attempt has been made by increasing the capacity of the UPFC from 100MW to 180MW [9] in step increase of 20MW to determine the System Indices. The system indices are tabulated in Table 4.8 and represented in Figs. 4.18 to 4.20 respectively for these indices.

Table 4.8: System Indices for Modified 6 bus RBTS at Generation Capacity of 240MW & Load Demand of 185MW – Original Scheme

UPFC Capacity	BPSD	BPII	Severity Index
100	19.834	0.364	329.642
120	18.641	0.361	327.916
140	17.319	0.355	326.441
160	15.911	0.349	323.78
180	14.164	0.342	321.661

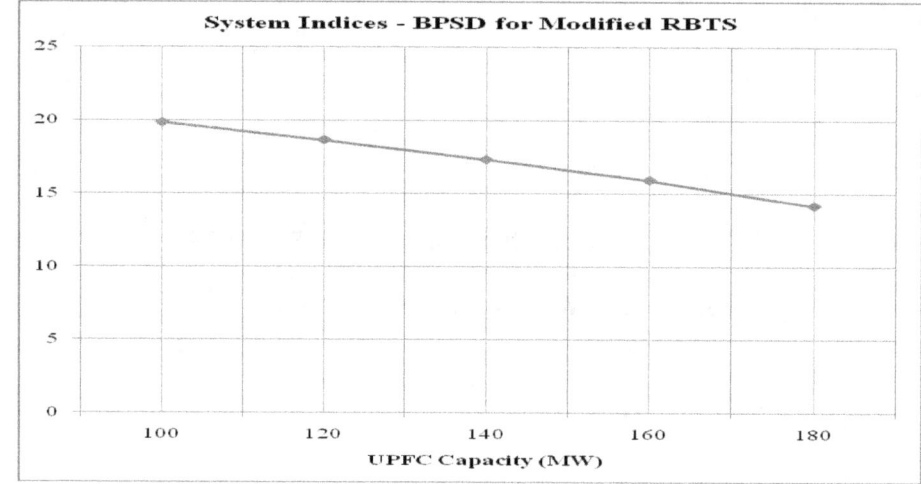

Fig. 4.18: BPSD vs UPFC Capacity of 6 Bus Modified RBTS

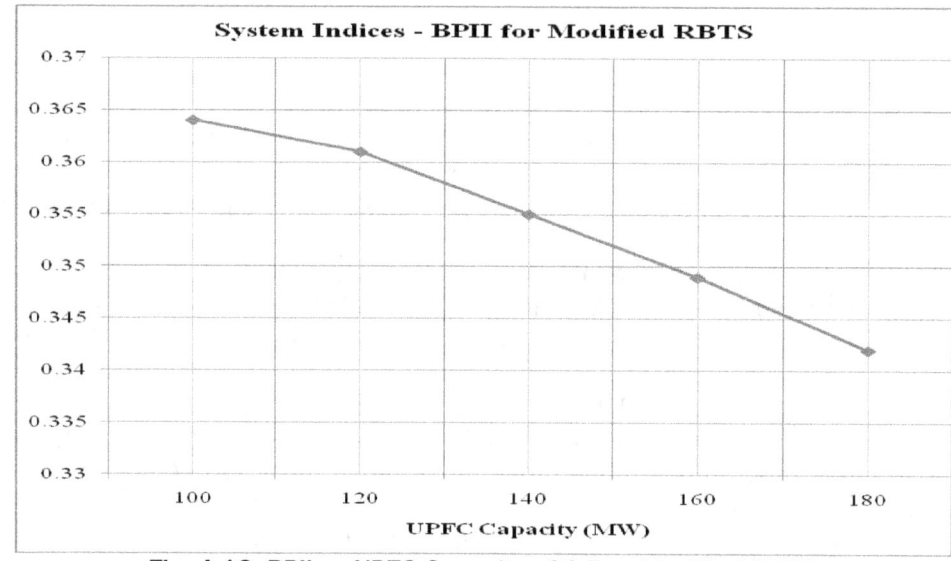

Fig. 4.19: BPII vs UPFC Capacity of 6 Bus Modified RBTS

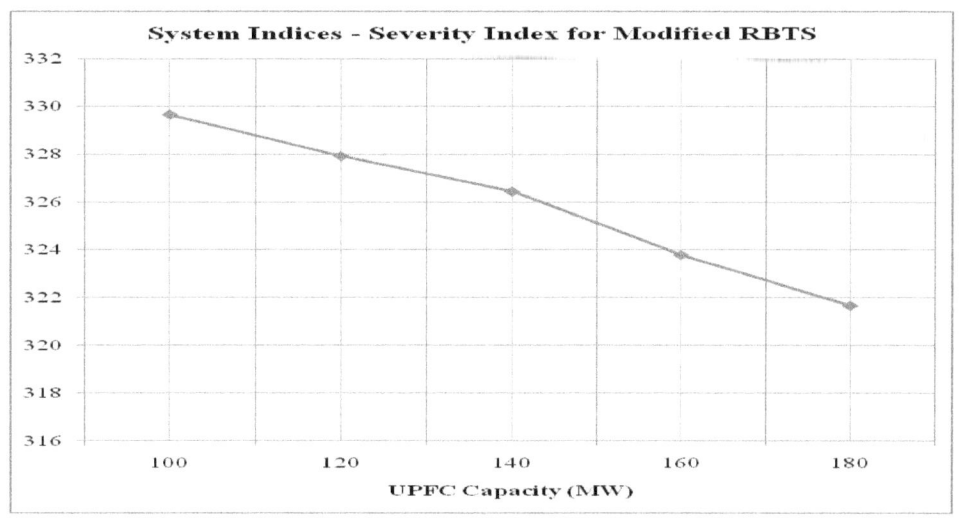

Fig. 4.20: BPECI vs UPFC Capacity of 6 Bus Modified RBTS

From Table 4.8, it can be noted that as the UPFC capacity of a single module is increased the system indices decreases. The graphical representation of Table 4.8 is shown in Figs. 4.18 to 4.20.

In order to determine higher efficiency of the system, the 6 bus RBTS is modified by increasing the capacity of UPFC from 100MW to 180MW in steps of 20MW. It can be observed from Fig. 4.18, 4.19 & 4.20 as the UPFC capacity is increasing the supply disturbance, Interruption Index and severity Index are decreasing since Hunt compensation & the control circuit of UPFC is enhanced as per the UPFC capacity.

Similarly the system Indices are calculated for the modified 6 Bus RBTS by considering different modules of UPFC and compared with the original scheme of the system. The compared system indices are tabulated in Tables 4.9, 4.10 & 4.11 and represented in Figs. 4.21, 4.22 & 4.23 respectively for these indices.

Table 4.9: BPSD for Modified 6 bus RBTS vs Number of Modules

UPFC Capacity	Number of Modules							
	Original	2	3	4	5	6	7	8
100 MW	19.834	19.11	18.53	17.62	16.57	15.45	14.96	14.62
120 MW	18.641	18.11	17.94	17.02	16.32	15.11	14.34	13.91
140 MW	17.319	16.94	16.54	16.11	15.29	14.88	14.24	13.77
160 MW	15.911	15.58	15.11	15.02	14.55	13.98	13.21	12.6
180 MW	14.164	13.88	13.37	12.97	12.01	11.49	10.99	10.45

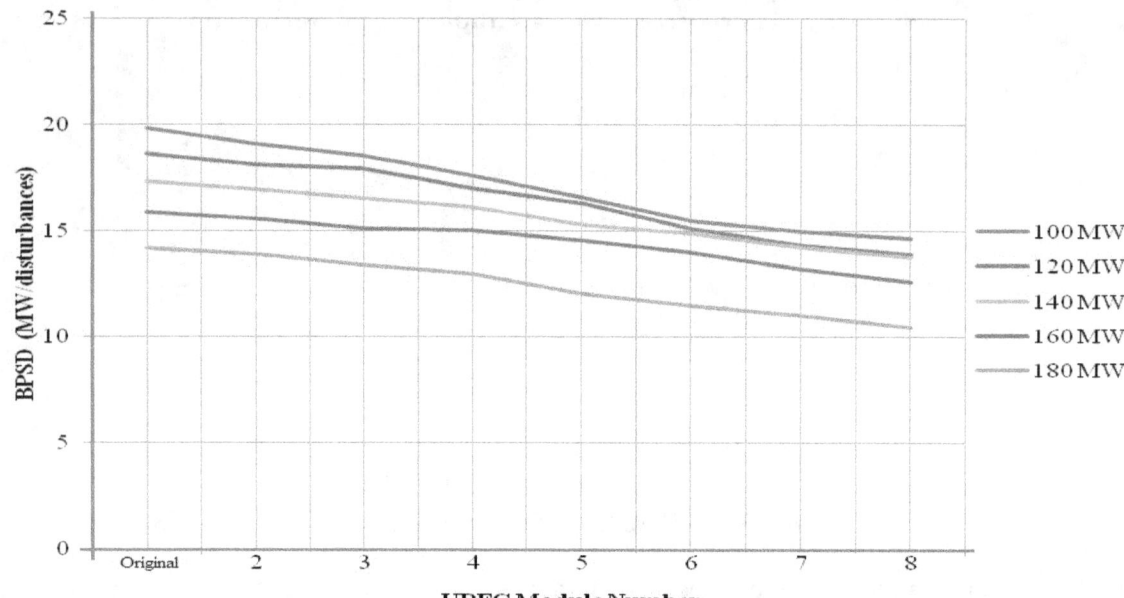

Fig. 4.21: BPSD vs UPFC Module Number of 6 Bus Modified RBTS

As the number of modules is increasing, the availability of the system is decreasing. In order to overcome this problem, it is proposed to increase the capacity of the UPFC component to attain the required load demand and the number of modules can be reduced which leads to increase in the availability of the system.

From Table 4.9, it can be observed that, as the UPFC capacity is increasing for different modules, the Bulk Power Supply Disturbance decreases which improves the overall performance of the system. Fig. 4.21 represents the graphical representation of Table 4.9.

Table 4.10: BPII for Modified 6 bus RBTS vs Number of Modules

UPFC Capacity	Number of Modules							
	Original	2	3	4	5	6	7	8
100 MW	0.364	0.362	0.359	0.357	0.356	0.353	0.351	0.35
120 MW	0.361	0.358	0.356	0.354	0.353	0.35	0.348	0.346
140 MW	0.355	0.353	0.35	0.347	0.345	0.342	0.34	0.339
160 MW	0.349	0.347	0.345	0.342	0.34	0.338	0.335	0.332
180 MW	0.342	0.34	0.338	0.336	0.333	0.331	0.329	0.327

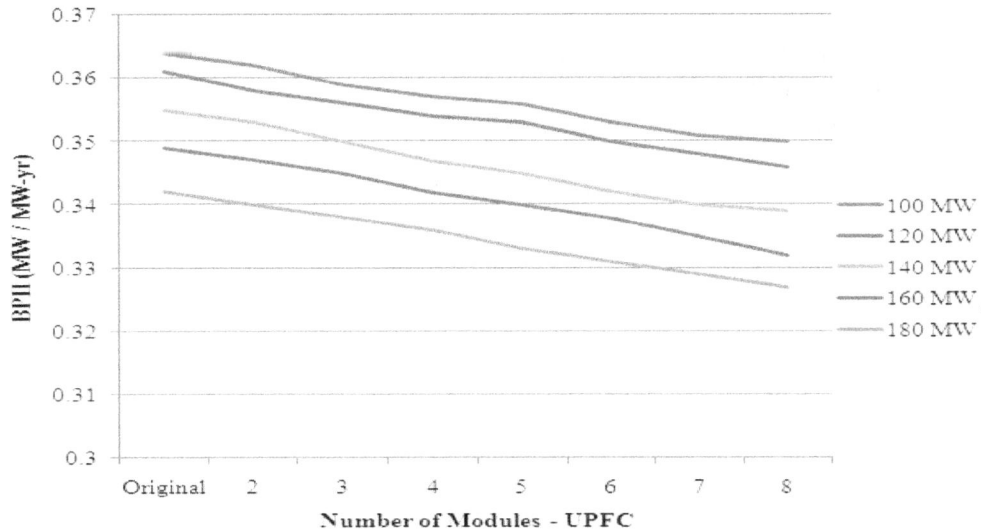

Fig. 4.22: BPII vs UPFC Module Number of 6 Bus Modified RBTS

From Table 4.10, it can be observed that, as the UPFC capacity is increasing for different modules, the Bulk Power Interruption Index decreases which improve the overall performance of the system. Fig. 4.22 represents the graphical form of Table 4.10.

Table 4.11: BPECI for Modified 6 bus RBTS vs Number of Modules

UPFC Capacity	Number of Modules							
	Original	2	3	4	5	6	7	8
100 MW	329.642	327.55	326.41	325.31	324.46	323.11	321.87	320.64
120 MW	327.916	326.77	325.47	324.51	323.09	322.47	321.38	320.05
140 MW	326.441	325.74	324.84	323.33	322.24	321.54	320.73	319.76
160 MW	323.78	322.66	321.84	320.55	319.67	318.44	317.99	317.55
180 MW	321.661	320.64	319.97	319.21	318.34	317.66	317.02	316.88

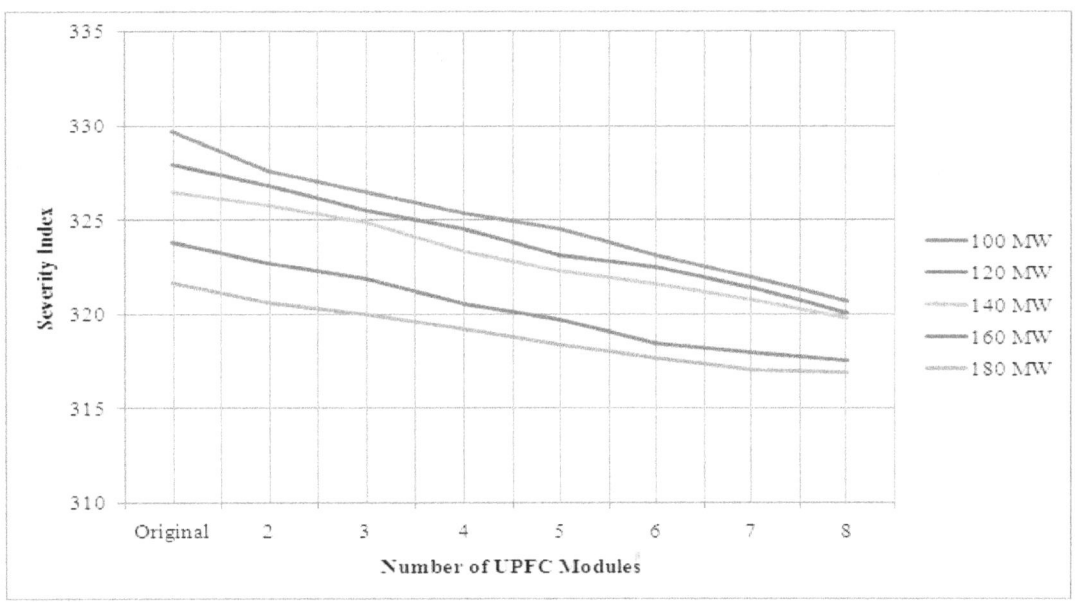

Fig. 4.23: BPECI vs UPFC Module Number of 6 Bus Modified RBTS

From Table 4.11, it can be observed that, as the UPFC capacity is increasing for different modules, the BPECI decreases which improves the overall performance of the system. Fig. 4.23 represents the graphical representation of Table 4.9.

For the 6 bus RBTS system the UPFC capacity are varied from 100MW to 180MW in steps of 20MW, simultaneously the number of modules used in the system are also varied to determine the supply disturbance, interruption index and severity index. From the analysis we can determine the optimal condition of the UPFC to be incorporated in the system.

Form Fig. 4.21, 4.22 & 4.23 it can be observed that as the UPFC capacity & number of modules are increased, the system indices are decreasing by 8% on an average.

4.3.3 Probability of Failure & EENS

The basic expected energy curtailed concept can also be used to determine the expected energy [11] produced by each unit in the system and therefore provides a relatively simple approach to production cost modeling.

Probability of failure = $Q_K = \sum_j P_j * P_{kj}$ . . . (4.15)

Where P_j = Probability of existence of outage j

P_{kj} = Probability of the load at bus K exceeding the maximum load that can be supplied at that bus during the outage j.

Expected Energy Not Supplied = $\sum_j L_{kj} * P_j * 8760 (MWh)$ . . . (4.16)

where L_{kj} = Load curtailment at bus K to alleviate line overloads arising due to the contingency j.

Probability of failure & EENS [6] are calculated for a 6 bus RBTS at each and every bus in the system and tabulated in Table 4.6 and represented in Figs. 4.14 & 4.15 respectively.

At Bus 2:

Probability of failure = Q_K = 0.9551663 * 0.0079579 = 0.00826652

Expected Energy Not Supplied = 0.01005 * 8760 = 88.082 (MWh)

Table 4.12: Probability of Failure & EENS of 6 bus RBTS

Bus No.	Probability of Failure	EENS
1	0.0081547	124.64
2	0.0082665	88.082
3	0.0083131	377.731
4	0.0083139	177.28
5	0.0083145	88.86
6	0.0084512	288.36

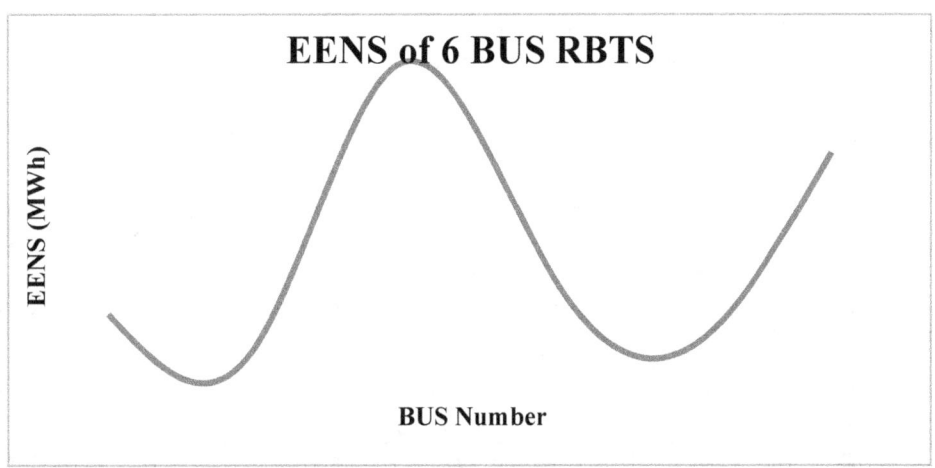

Fig. 4.24: Probability of Failure vs Bus Number of 6 Bus RBTS

EENS of 6 BUS RBTS

Fig. 4.25: EENS vs Bus Number of 6 Bus RBTS

From Table 4.12, it can be noted that, although EENS at bus 3 is higher rather than other buses, the probability of failure is less when compared with bus 6. Similarly, EENS at bus 6 is less than that of bus 3. Figs. 4.24 & 4.25 represent the graphical form of Table. 4.12.

Similarly the probability of failure & EENS for 6 Bus RBTS is calculated with different modules of UPFC which are tabulated in Tables 4.13 & 4.14 and represented in Figs. 4.26 & 4.27 respectively.

Table 4.13: Probability of Failure for 6 bus RBTS with different Modules of UPFC

Module No.	Bus No.					
	1	2	3	4	5	6
2	0.008232	0.008363	0.008141	0.008137	0.008142	0.00876
3	0.008215	0.008348	0.008167	0.008162	0.008171	0.00886
4	0.008194	0.008331	0.008248	0.008245	0.008252	0.00894
5	0.008179	0.008316	0.008281	0.008275	0.008281	0.00907
6	0.008158	0.008292	0.008307	0.008306	0.008307	0.00925
7	0.008154	0.008266	0.008313	0.008313	0.008314	0.00945
8	0.008162	0.008297	0.008301	0.008302	0.008306	0.00916

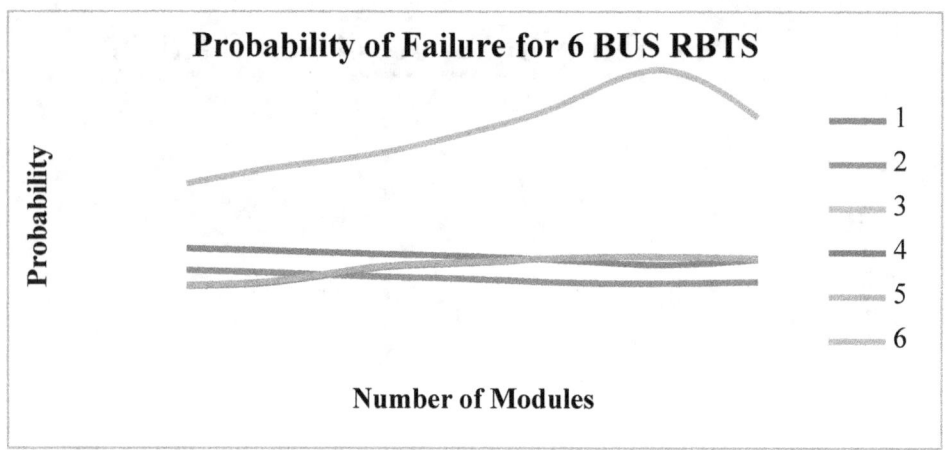

Fig. 4.26: Probability of Failure vs Number of Modules of UPFC for 6 Bus RBTS

From Table 4.13, it can be observed that the Probability of Failure is decreasing as the number of modules increases at each bus. As discussed earlier due to thermal limit, 8 modules UPFC is not suitable for the system that is why the Probability of Failure increases at 8th Module. The graphical form of Table. 4.13 is shown in Fig. 4.26.

Table 4.14: EENS for 6 bus RBTS with different Modules of UPFC

Module No.	Bus No.					
	1	2	3	4	5	6
2	126.94	90.02	381.21	179.66	91.48	292.35
3	126.21	89.21	379.78	178.38	90.68	290.34
4	125.94	88.81	378.94	177.97	90.31	289.97
5	125.54	88.39	378.25	177.69	89.81	289.34
6	125.12	88.12	377.86	177.51	89.42	288.94
7	124.64	88.08	377.73	177.28	88.86	288.36
8	125.21	88.91	378.14	178.85	89.81	289.02

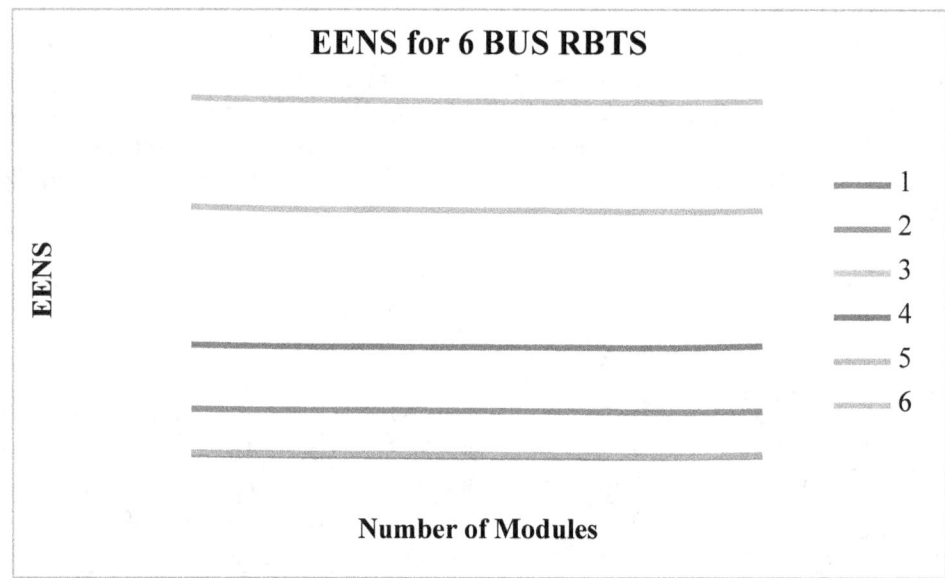

Fig. 4.27: EENS vs Number of Modules of UPFC for 6 Bus RBTS

From Table 4.14, it can be observed that the EENS is decreasing as the number of modules increases at each bus. As discussed earlier due to thermal limit, 8 modules UPFC is not suitable for the system that is why the EENS increases at 8th Module. The graphical representation of Table. 4.13 is shown in Fig. 4.26.

Fig. 4.26 & 4.27 shows the probability of failure & EENS for a 6 bus RBTS for different module UPFC at each bus of the system. It can also be observed that the probability of failure is decreasing at each bus as the number of modules is increasing. As, it has been already decided to use 7 module UPFC for the system, as the number of modules are exceeding the desired value the probability of failure & EENS are increasing.

4.4 Conclusions

In this chapter, the analysis of 6 bus RBTS is determined when using UPFC with different modules in all the bus. From the above results, depending upon the generation & transmission capacity it can be concluded that seven module UPFC is suitable for the system based on the reliability of different modules. Apart from the reliability, system indices, probability of failure & EENS shows a major improvement in the system of different modules for all the buses.

Chapter 5

RELIABILITY ANALYSIS OF COMPOSITE POWER SYSTEM USING TCSC

5.1 Introduction

TCSC has been demonstrated to offer a variety of benefits to power systems. It can alter its reactance rapidly and smoothly, in turn the apparent impedance of the corresponding transmission line. Therefore, a TCSC can change the power flow of the network. If the parameters of TCSC are set properly, power might flow along the desired path. That means the system may operate in a better condition with the employment of a TCSC. Hence, if contingency happens the loss of load or energy will be expressed to reduce when a TCSC is installed in the grid. The single line diagram of one module with TCSC is shown in Fig. 5.1.

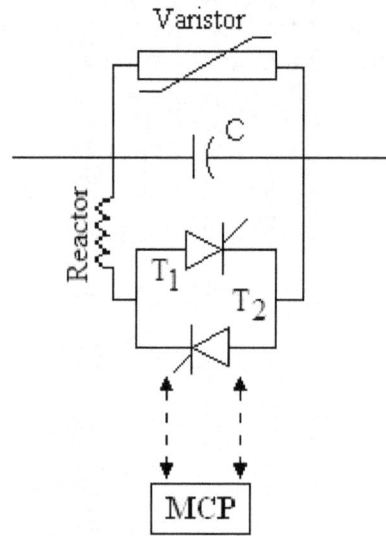

MCP – Module Control & Protection unit
Fig. 5.1: Single line diagram of TCSC (One Module)

In this Chapter, the reliability analysis of composite power system using TCSC is presented. The reliability analysis viz. Availability and Unavailability for multi-module TCSC is discussed by using State Space representation in Section 5.2. Also, the System Indices BPSD, BPII and BPECI, Probability of Failure and EENS of IEEE 6 Bus Roy Billinton Test System (RBTS) are discussed with all results. Conclusions of Chapter 4 are presented in Section 5.3.

5.2 Reliability Analysis of TCSC

The availability and reliability of a TCSC depends on the performance of each element in the TCSC. For a multi module TCSC, two different types of failure modes can be assigned to each module. The cause of outage in only the module itself refers to first cause eg. capacitor failure, the cause of outage of all M-modules refers to second cause eg. Varistor failure.

The reliability analysis of TCSC is determined by the state-space model. In the state space model, only the first failure mode is considered to cause the module to be transferred to the down state. Based on the two state model, once a module is declared failed, it is not available for line compensation. The steady state probabilities P_1 & P_2 for the up & down states are [78]:

$$P_1 = \frac{\mu}{\mu + \lambda} \quad \quad \ldots \quad (5.1)$$

$$P_2 = \frac{\lambda}{\mu + \lambda} \qquad (5.2)$$

where λ is Failure Rate and μ is Repair Rate.

By considering different modules of TCSC for line compensation the availability & unavailability can be determined by using state space representation as discussed in Chapter 4. In Table 5.1, the availability & unavailability of different module TCSC are presented.

Table 5.1: Availability & Unavailability of different Modules - TCSC

Modules	Availability	Unavailability
2	0.995371	0.004629
3	0.994594	0.005406
4	0.993628	0.006372
5	0.992594	0.007406
6	0.991834	0.008166
7	0.990346	0.009654
8	0.980671	0.019329

From Table 5.1, it can be observed that as number of modules increases the availability of the system decreases.

5.2.1 System Indices

System indices BPSD, BPII & BPECI are calculated for 6 bus RBTS by incorporating TCSC in the system by replacing UPFC. The system Indices can be found out from Eqns. (4.12), (4.13) & (4.14) respectively.

The system indices for a 6 Bus RBTS are calculated by changing the number of modules of TCSC and are tabulated in Table 5.2 and represented in Figs. 5.2, 5.3 & 5.4 respectively.

Table 5.2: System Indices of 6 Bus RBTS with different modules of TCSC

No. of Modules	BPSD	BPII	BPECI
2	17.54	0.3462	284.36
3	17.17	0.3458	284.01
4	16.85	0.3451	283.92
5	16.58	0.345	283.87
6	16.51	0.3448	283.85
7	16.62	0.3452	283.88
8	16.84	0.3456	283.91

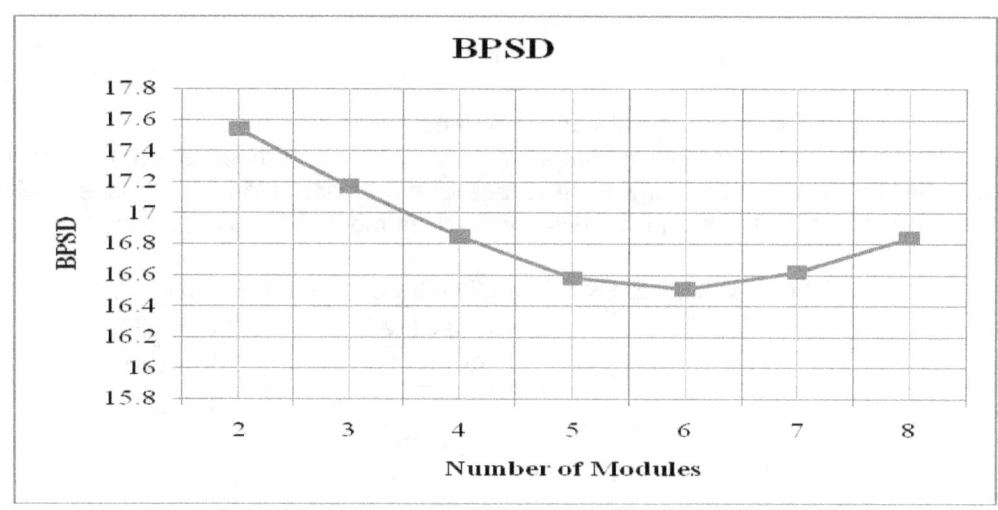

Fig. 5.2: BPSD vs Number of Modules of TCSC in 6 Bus RBTS

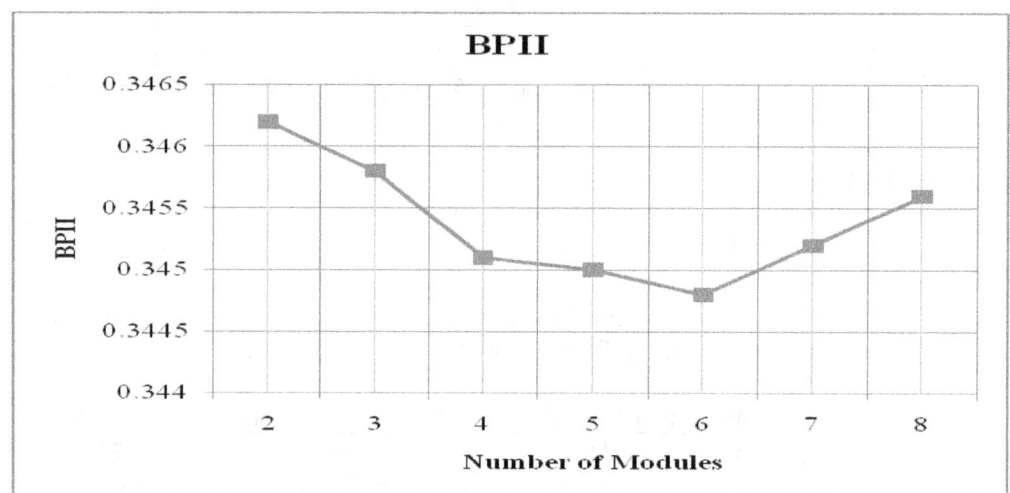

Fig. 5.3: BPII vs Number of Modules of TCSC in 6 Bus RBTS

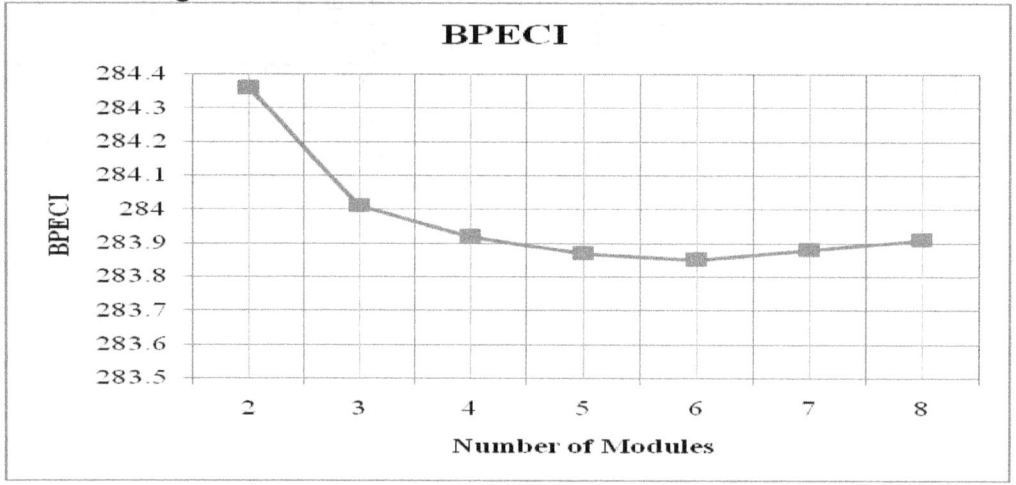

Fig. 5.4: BPECI vs Number of Modules of TCSC in 6 Bus RBTS

From Table 5.2, it can be observed that the system indices BPSD, BPII and BPECI are decreasing as the number of modules increasing. For the system considered the number of modules required is Seven. As the number of modules increases beyond the required number the system indices are also increasing, and this is due to over thermal loading of FACTS devices beyond their limit, which can be observed from Table 5.2. The graphical representation of Table 5.2 can be observed in Figs. 5.2 to 5.4.

Systems indices BPSD, BPII & BPECI are calculated with respect to the generation capacity (MW) by considering 7 modules TCSC for 6 Bus RBTS. The system indices are tabulated in Table 5.3. A comparison has been made when using 7 modules TCSC & without using TCSC is presented in Figs. 5.5, 5.6 & 5.7 respectively.

Table 5.3: System Indices vs Generation Capacity – with 7 modules TCSC

Generation Capacity (MW)	Load Demand (MW)	BPSD	BPII	BPECI
240	185	14.35	0.246	259.86
270	203.5	14.57	0.261	263.66
300	222	15.02	0.269	268.71
330	240.5	16.72	0.283	271.93
345	259	20.63	0.315	288.32
360	277.5	28.146	0.364	297.64

From Table 5.3, it can be observed that, system indices BPSD, BPII & BPECI increases as the Generation capacity and Load Demand of the system increases, without using any FACTS controlling devices in the given system. This is because, load at the given bus or transmission line cannot be predicted directly. The graphical form of Table 5.3 is shown in Figs. 5.5 to 5.7.

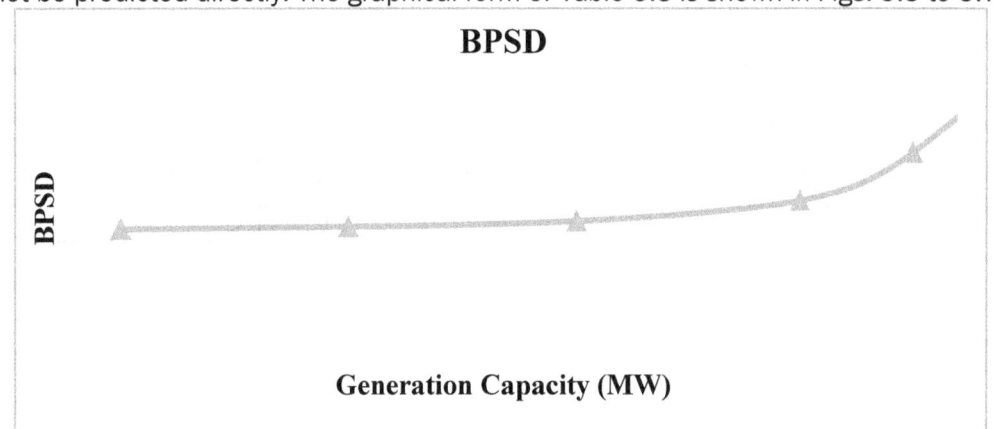

Fig. 5.5: BPSD vs Generation Capacity of 6 Bus RBTS

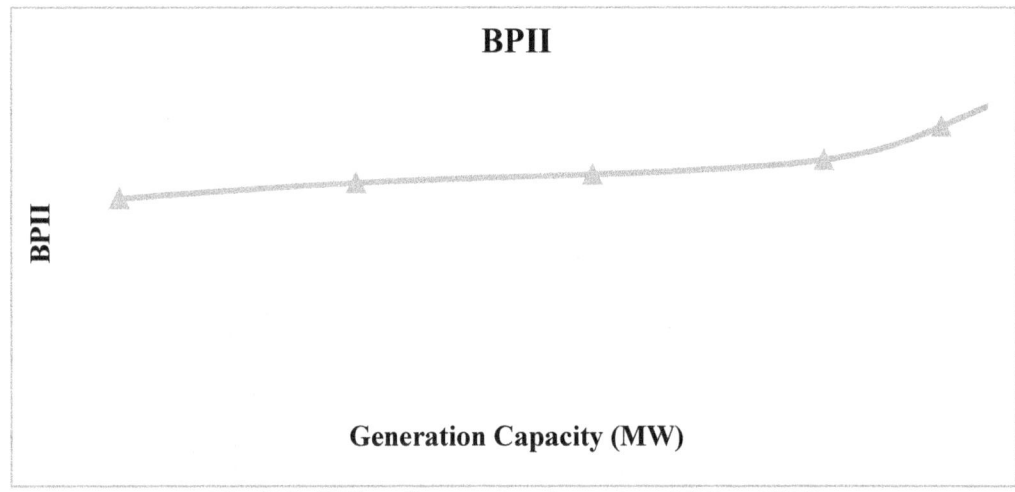

Fig. 5.6: BPII vs Generation Capacity of 6 Bus RBTS

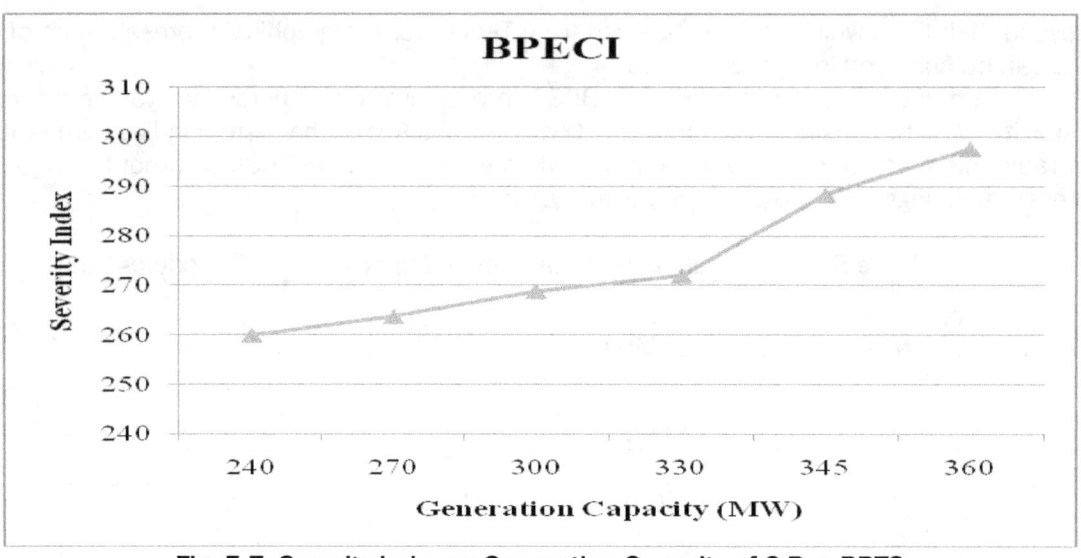

Fig. 5.7: Severity Index vs Generation Capacity of 6 Bus RBTS

Similarly as of UPFC, an attempt has been made by increasing the capacity of the TCSC from 100MW to 180MW in step increase of 20MW to determine the System Indices. The system indices are tabulated in Table 5.4 and represented in Figs. 5.8, 5.9 & 5.10 respectively.

Table 5.4: System Indices for Modified 6 bus RBTS at Generation Capacity of 240MW

TCSC Capacity (MW)	BPSD	BPII	BPECI
100	39.98	0.562	389.16
120	30.62	0.432	343.79
140	28.36	0.416	342.11
160	28.11	0.408	341.86
180	27.86	0.405	340.12

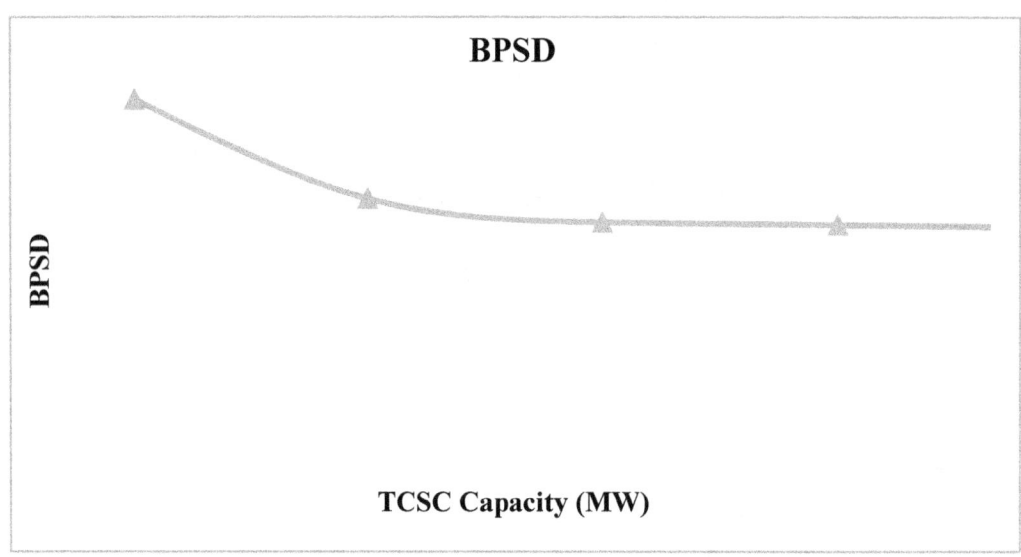

Fig. 5.8: BPSD vs TCSC Capacity of 6 Bus Modified RBTS

From Table 5.4 it can be observed that, as the TCSC capacity is increasing for different modules the system indices BPSD, BPII and BPECI decreases which improves the overall performance of the system. Figs. 5.8, 5.9 & 5.10 represents the graphical representation of Table 5.4.

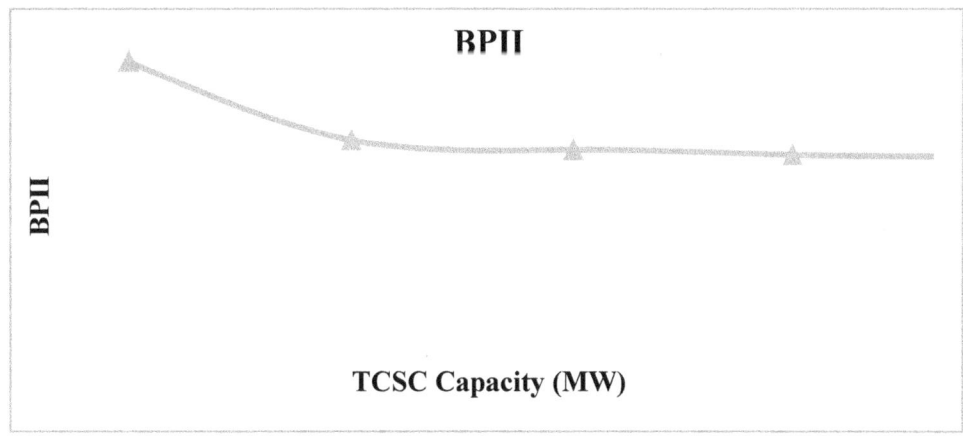

Fig. 5.9: BPII vs TCSC Capacity of 6 Bus Modified RBTS

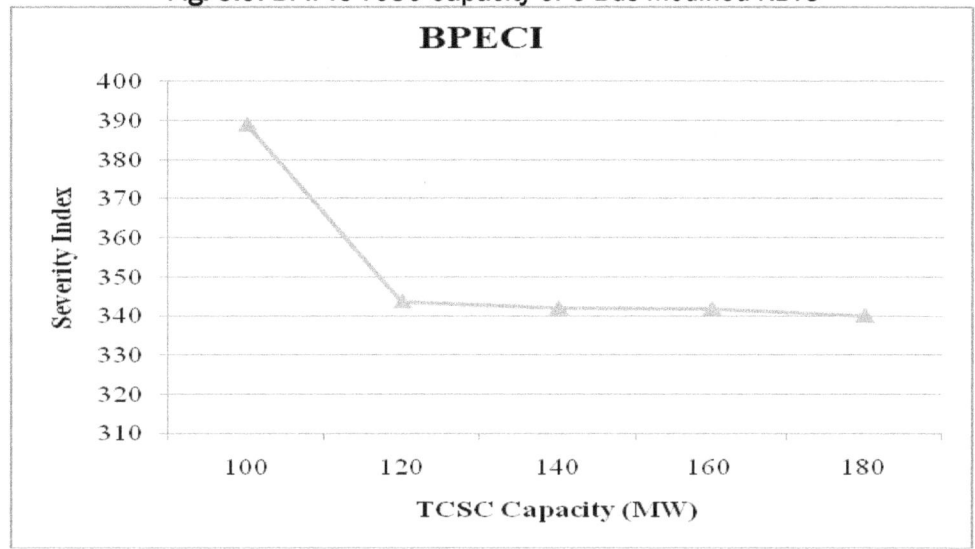

Fig. 5.10: BPECI vs TCSC Capacity of 6 Bus Modified RBTS

5.2.2 Probability of Failure & EENS

Probability of failure & EENS are calculated for a 6 bus RBTS at each and every bus in the system and tabulated in Table 5.5 and represented in Figs. 5.11 & 5.12 respectively.

At Bus 2:

Probability of failure = Q_K = 0.993628 * 0.00851727 = 0.008463   using Eqn. (4.15).

Expected Energy Not Supplied = 0.993628 * 0.011486 * 8760
= 0.011413 * 8760 = 99.98 (MWh)   using Eqn. (4.16).

Table 5.5: Probability of Failure & EENS of 6 bus RBTS

Bus No.	Probability of Failure	EENS
1	0.008411	129.94
2	0.008463	99.98
3	0.008458	395.43
4	0.008456	183.94
5	0.008457	97.58
6	0.008472	298.66

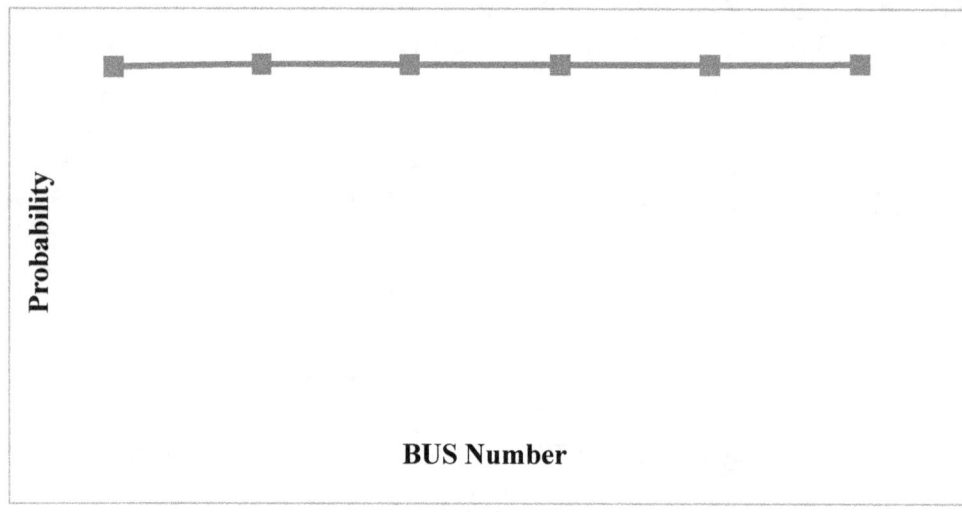

Fig. 5.11: Probability of Failure vs Bus Number of 6 Bus RBTS

From Table 5.5, it can be noted that, although EENS at bus 3 is higher rather than other buses, the probability of failure is less when compared with bus 6. Similarly, EENS at bus 6 is less than that of bus 3. Figs. 5.11 & 5.12 represent the graphical form of Table. 5.5.

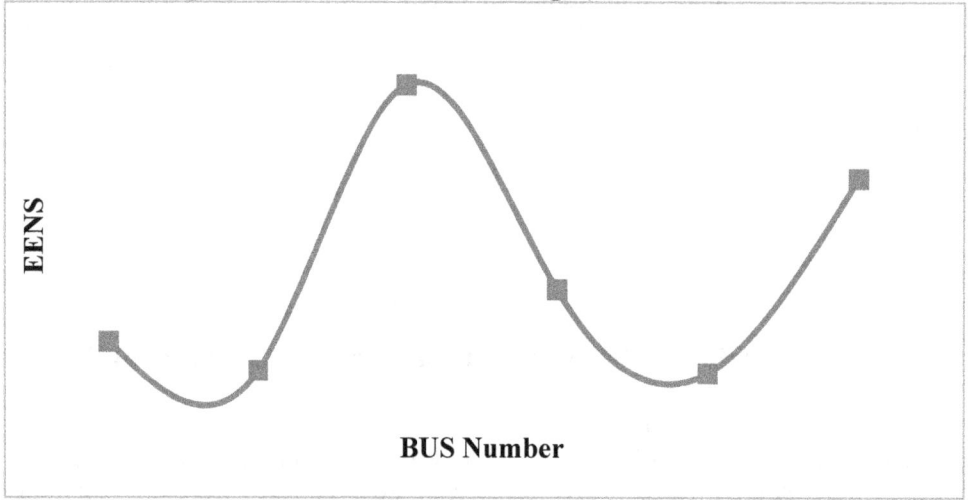

Fig. 5.12: EENS vs Bus Number of 6 Bus RBTS

Similarly the probability of failure & EENS for 6 Bus RBTS is calculated with different modules of TCSC which are tabulated in Tables 5.6 & 5.7 and represented in Figs. 5.13 & 5.14 respectively.

Table 5.6: Probability of Failure for 6 bus RBTS with different Modules of TCSC

Module No.	Bus No.					
	1	2	3	4	5	6
2	0.008536	0.008516	0.008513	0.008511	0.008507	0.008634
3	0.008511	0.008507	0.008501	0.008498	0.008496	0.008602
4	0.008459	0.008498	0.008489	0.008486	0.008489	0.008562
5	0.008416	0.008491	0.008481	0.008479	0.008480	0.008511
6	0.008411	0.008463	0.008458	0.008456	0.008457	0.008472
7	0.008417	0.008475	0.008474	0.008472	0.008470	0.008489
8	0.008426	0.008494	0.008496	0.008494	0.008491	0.008496

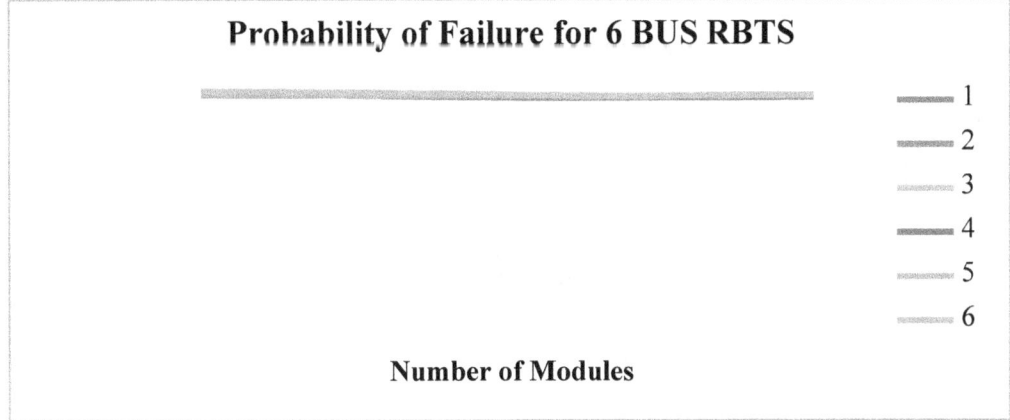

Fig. 5.13: Probability of Failure vs Number of Modules of TCSC for 6 Bus RBTS

From Table 5.6, it can be observed that the Probability of Failure is decreasing as the number of modules increases at each bus. As discussed earlier due to thermal limit, 8 modules UPFC is not suitable for the system that is why the Probability of Failure increases at 8th Module. The graphical representation of Table. 5.6 is shown in Fig. 5.13.

Table 5.7: EENS for 6 bus RBTS with different Modules of TCSC

Module No.	Bus No.					
	1	2	3	4	5	6
2	132.64	102.34	398.64	186.42	100.24	302.84
3	132.12	101.96	398.14	185.96	99.38	301.63
4	131.54	101.12	397.58	185.13	98.66	300.75
5	131.08	100.78	396.91	184.66	97.81	299.54
6	129.94	99.98	395.43	183.94	97.58	298.66
7	130.16	100.36	396.16	184.52	98.12	299.18
8	130.84	101.03	396.88	185.66	99.42	300.11

Fig. 5.14: EENS vs Number of Modules of TCSC for 6 Bus RBTS

From Table 5.7, it can be observed that the EENS is decreasing as the number of modules increases at each bus. As discussed earlier due to thermal limit, 8 modules UPFC is not suitable for the system that is why the EENS increases at 8th Module. The graphical representation of Table. 5.7 is shown in Fig. 5.14.

5.3 Conclusion

In this chapter, the analysis of 6 bus RBTS is determined when using TCSC with different modules in all the bus. From the above results, it can be concluded that seven module TCSC is suitable for the system based on the reliability of different modules depending upon the generation & transmission capacity. Apart from the reliability, system indices, probability of failure & EENS shows a major improvement in the system of different modules for all the buses.

Chapter 6

RELIABILITY ANALYSIS OF COMPOSITE POWER SYSTEM USING TCSC, UPFC – A COMPARISON

6.1 Introduction

Reliability analysis of Composite Power System is determined by applying the FACTS devices like UPFC & TCSC in the system at different buses & transmission lines. In order, to describe which FACTS device is best suitable for the system, a comparison is carried out between the systems with UPFC, TCSC. The comparison is made in different parameters viz. availability & unavailability, System Indices, Probability of Failure & Expected Energy Not Supplied etc., by considering different modules & bus numbers of the composite power system.

TCSC is one of the most important and best known FACTS devices, which has been in use for many years to increase the power transfer as well as to enhance system stability. The TCSC consists of three main components: capacitor bank C, bypass inductor L and bidirectional thyristors SCR1 and SCR2. The firing angles of the thyristors are controlled to adjust the TCSC reactance in accordance with a system control algorithm, normally in response to some system parameter variations. According to the variation of the thyristor firing angle or conduction angle, this process can be modeled as a fast switch between corresponding reactances offered to the power system.

Unified Power Flow Controller (UPFC) is the most versatile one that can be used to enhance steady state stability, dynamic stability and transient stability. The UPFC is capable of both supplying and absorbing real and reactive power and it consists of two ac/dc converters. One of the two converters is connected in series with the transmission line through a series transformer and the other in parallel with the line through a shunt transformer. The DC side of the two converters is connected through a common capacitor, which provides dc voltage for the converter operation. The power balance between the series and shunt converters is a prerequisite to maintain a constant voltage across the DC capacitor. As the series branch of the UPFC injects a voltage of variable magnitude and phase angle, it can exchange real power with the transmission line and thus improves the power flow capability of the line as well as its transient stability limit. The shunt converter exchanges a current of controllable magnitude and power factor angle with the power system. It is normally controlled to balance the real power absorbed from or injected into the power system by the series converter plus the losses by regulating the dc bus voltage at a desired value.

In this Chapter, comparison for the reliability analysis of composite power system using TCSC, UPFC is presented. Comparison of reliability analysis with respect to Availability & Unavailability of the system is presented in Section 6.2. Comparison of reliability analysis with respect to system indices viz. BPSD, BPII and BPECI of the system is presented in Section 6.3. In Section 6.4, Comparison of reliability analysis with respect to Probability of Failure & EENS of the system is presented. Conclusions of Chapter 6 are presented in Section 6.5.

6.2 Comparison with respect to Availability & Unavailability

Availability & unavailability of the system when using UPFC, TCSC are calculated individually in Chapters 4 and 5 respectively. A comparison is made between the two FACTS elements of the system which is shown in Table 6.1 when using different modules. The graphical representation of Table 6.1 is shown in Figs. 6.1 & 6.2 respectively.

Table 6.1: Availability & Unavailability of UPFC, TCSC with different Modules

Modules	Availability		Unavailability	
	UPFC	TCSC	UPFC	TCSC
2	0.99818	0.995371	0.00182	0.004629
3	0.99623	0.994594	0.00377	0.005406
4	0.99462	0.993628	0.00538	0.006372
5	0.99374	0.992594	0.00626	0.007406
6	0.99246	0.991834	0.00754	0.008166
7	0.99164	0.990346	0.00836	0.009654
8	0.98254	0.980671	0.01746	0.019329

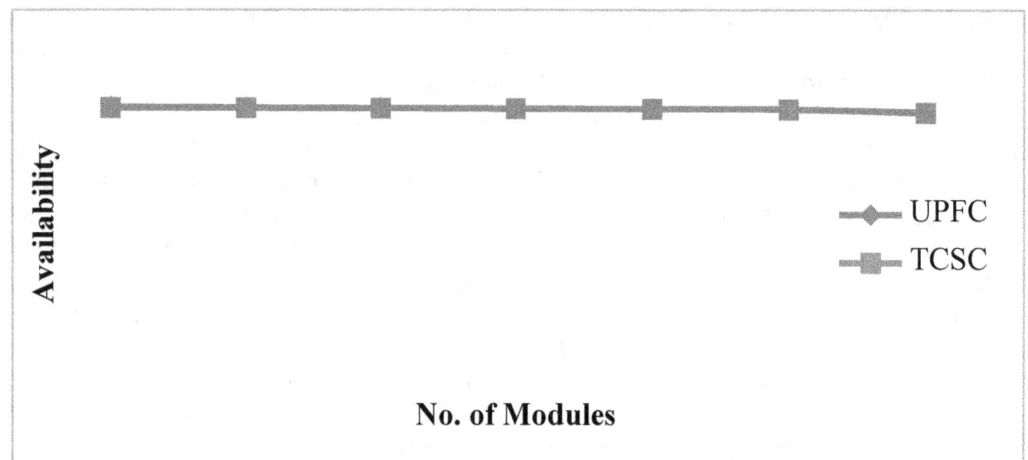

Fig. 6.1: Availability of FACTS devices vs No. of Modules

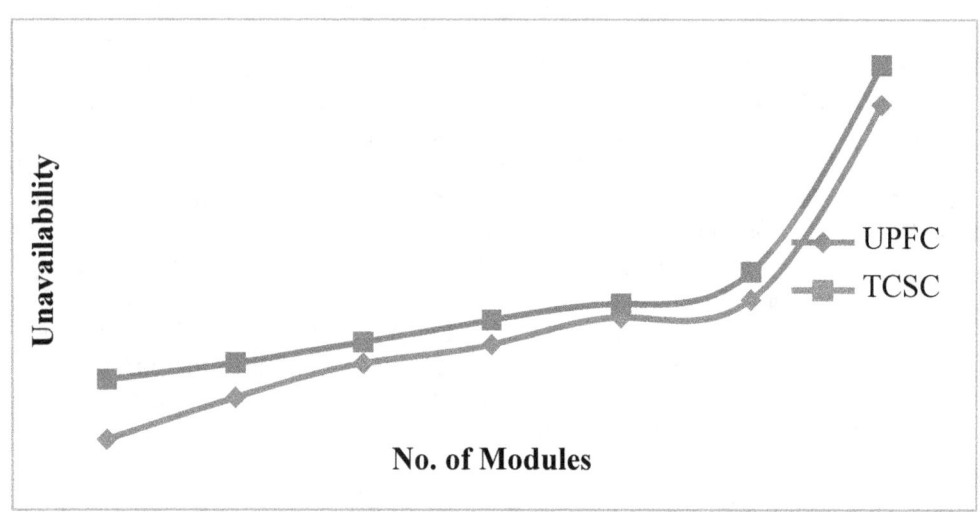

Fig. 6.2: Unavailability of FACTS devices vs No. of Modules

From Table 6.1, it can be observed that UPFC is having more availability rather than TCSC, or in other words, TCSC is having more unavailability rather than UPFC. The graphical representation of Table 6.1 is shown in Figs. 6.1 & 6.2 respectively.

6.3 Comparison with respect to System Indices

System Indices BPSD, BPII & BPECI for 6 bus RBTS are calculated with by applying different modules of FACTS devices which have been discussed already in Chapters 4 and 5. A comparison for system indices are determined in Tables 6.2, 6.3 & 6.4 when using UPFC, TCSC in the given system & demonstrated in Figs. 6.3, 6.4 & 6.5 respectively.

Table 6.2: Comparison of BPSD with different modules of UPFC & TCSC

Module. No.	TCSC	UPFC
2	19.69	17.54
3	19.58	17.17
4	19.47	16.85
5	19.39	16.58
6	19.31	16.51
7	19.26	16.62
8	19.38	16.84

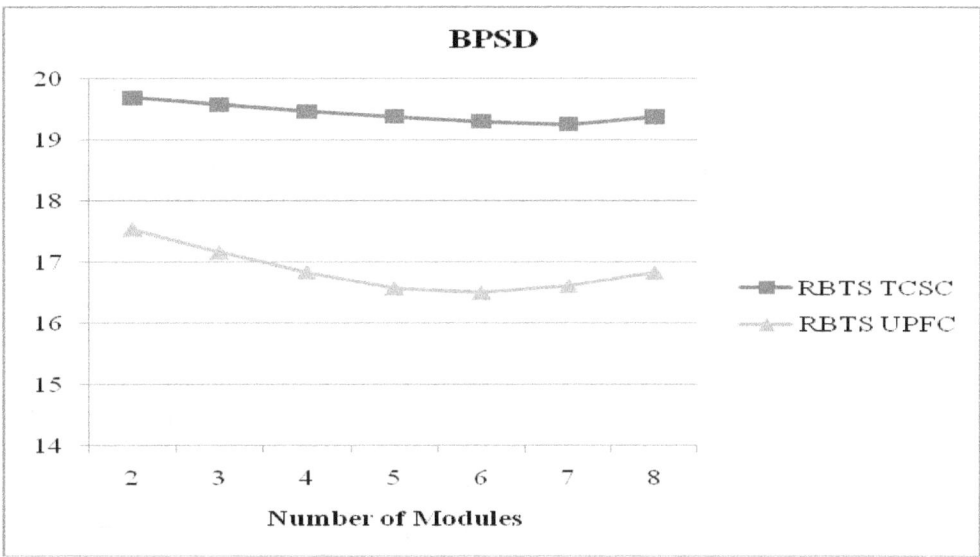

Fig 6.3: Comparison of BPSD with No. of Modules of UPFC & TCSC

From Table 6.2, it can be observed that Bulk Power Supply Disturbance is less when UPFC is incorporated into the system rather than of TCSC for change in number of modules. The graphical representation of Table 6.2 is shown in Fig. 6.3.

Table 6.3: Comparison of BPII with different modules of UPFC & TCSC

Module No.	TCSC	UPFC
2	0.3642	0.3462
3	0.363	0.3458
4	0.3622	0.3451
5	0.3612	0.345
6	0.361	0.3448
7	0.359	0.3452
8	0.364	0.3456

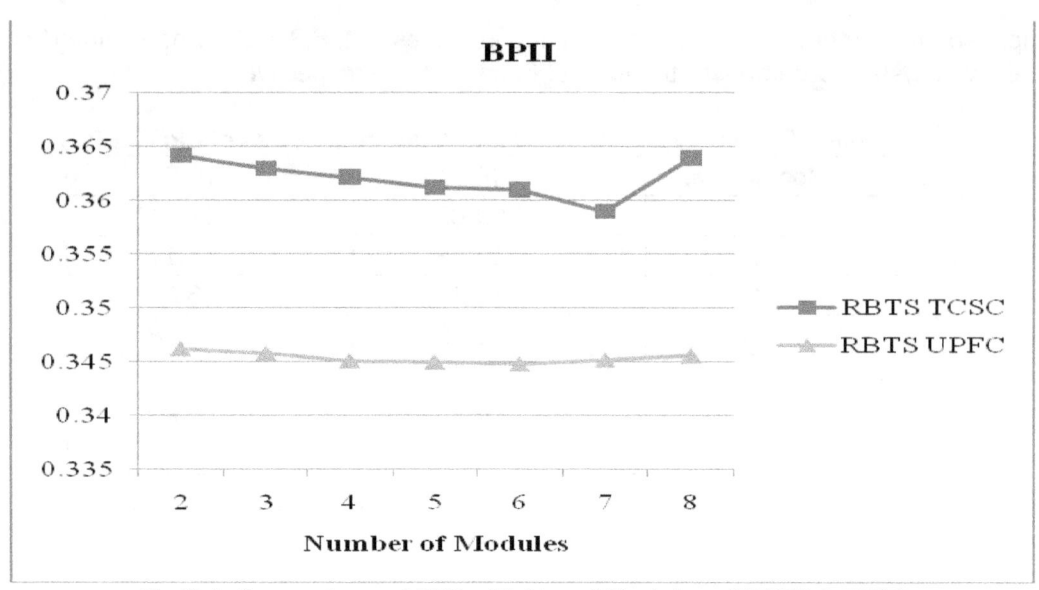

Fig 6.4: Comparison of BPII with No. of Modules of UPFC & TCSC

From Table 6.3, it can be observed that Bulk Power Interruption Index is less when UPFC is incorporated into the system rather than of TCSC for change in number of modules. The graphical representation of Table 6.3 is shown in Fig. 6.4.

Table 6.4: Comparison of BPECI with different modules of UPFC & TCSC

Module No.	TCSC	UPFC
2	331.14	284.36
3	331.11	284.01
4	331.08	283.92
5	331.04	283.87
6	330.91	283.85
7	330.69	283.88
8	331.01	283.91

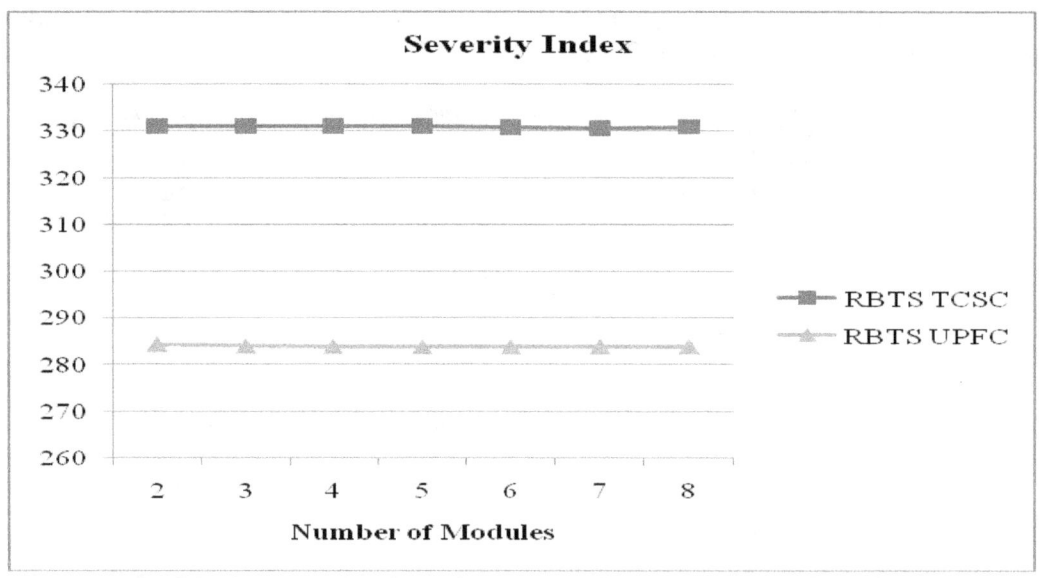

Fig 6.5: Comparison of BPECI with No. of Modules of UPFC & TCSC

From Table 6.4, it can be observed that Bulk Power Energy Curtailment Index is less when UPFC is incorporated into the system rather than of TCSC for change in number of modules. The graphical representation of Table 6.4 is shown in Fig. 6.5.

Similarly the system indices are also calculated with respect to the generation capacity, load demand when using UPFC, TCSC and when not using UPFC, TCSC. The corresponding values are represented in Tables 6.5 & 6.6 and graphically represented in Figs. 6.6, 6.7 & 6.8 respectively.

Table 6.5: System Indices (with & without UPFC) vs Generation Capacity

Generation Capacity (MW)	Load Demand (MW)	BPSD		BPII		BPECI	
		Without UPFC	With UPFC	Without UPFC	With UPFC	Without UPFC	With UPFC
240	185	14.76	14.35	0.274	0.246	264.66	259.86
270	203.5	14.98	14.57	0.286	0.261	272.34	263.66
300	222	15.63	15.02	0.297	0.269	276.93	268.71
330	240.5	17.24	16.72	0.308	0.283	278.11	271.93
345	259	24.11	20.63	0.395	0.315	294.63	288.32
360	277.5	68.73	28.146	0.632	0.364	563.66	297.64

From Table 6.5 it is clear that, the system indices like BPII, BPSD and BPECI are reducing when UPFC is incorporated into the system as compared with the same system without UPFC for different Generation capacity & Load demands. If the disturbances & losses are reduced, the efficiency of the system increases.

Table 6.6: System Indices (with & without TCSC) vs Generation Capacity

Generation Capacity (MW)	Load Demand (MW)	BPSD		BPII		BPECI	
		Without TCSC	With TCSC	Without TCSC	With TCSC	Without TCSC	With TCSC
240	185	19.834	19.428	0.364	0.324	329.642	328.521
270	203.5	19.897	19.679	0.372	0.334	331.246	330.125
300	222	20.653	20.315	0.379	0.351	333.201	331.921
330	240.5	21.842	21.242	0.389	0.360	336.721	332.671
345	259	34.197	23.157	0.465	0.378	371.264	333.164
360	277.5	78.673	29.736	0.913	0.389	615.259	334.259

From Table 6.6 it is clear that, the system indices like BPII, BPSD and BPECI are reducing when TCSC is incorporated into the system as compared with the same system without TCSC for different Generation capacity & Load demands. If the disturbances & losses are reduced, the efficiency of the system increases. The graphical representation of Tables 6.5 and 6.6 are shown in Figs. 6.6, 6.7 and 6.8 respectively.

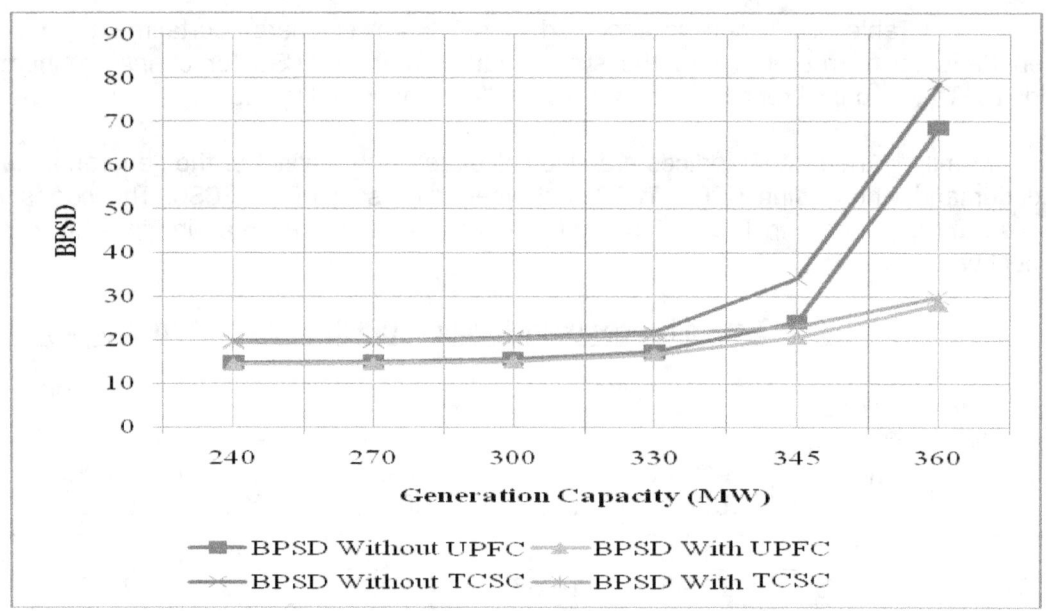

Fig 6.6: System Indices (BPSD) vs Generation Capacity

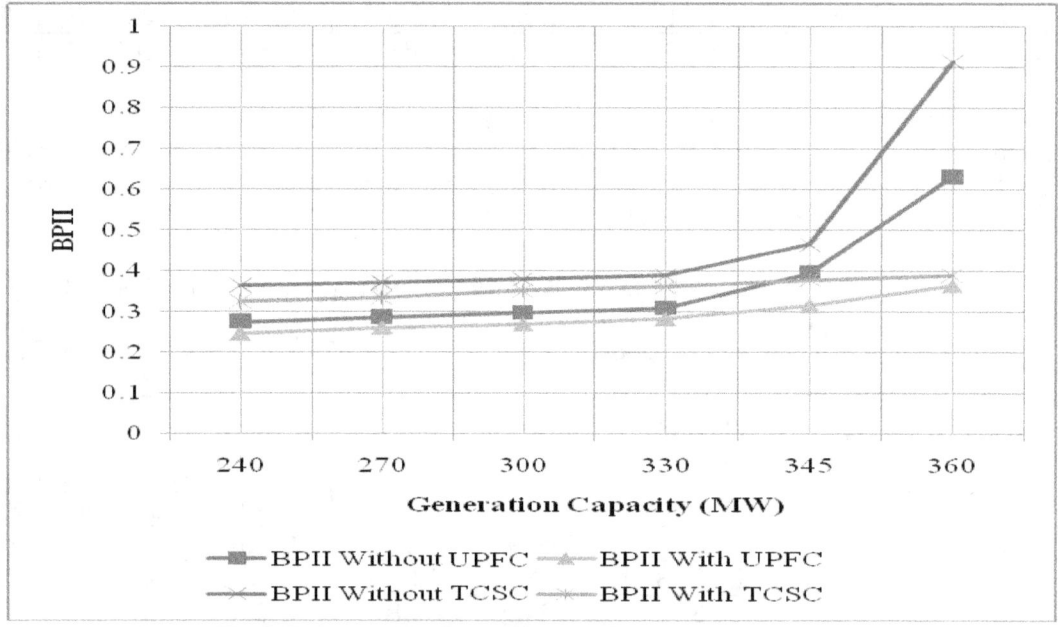

Fig 6.7: System Indices (BPII) vs Generation Capacity

From Fig. 6.7 it can be observed that as the Generating Capacity of the System is increasing the Bulk Power Interruption Index is also Increasing. When the system is not utilizing any FACTS devices (TCSC, UPFC) the interruption index are more as the generating capacity is increasing. Since the compensation in UPFC is greater than TCSC the interruption index is less in UPFC as compared with TCSC. From the graph it is clearly observed that the interruption index is less when the system is incorporated with UPFC when compared with the remaining conditions

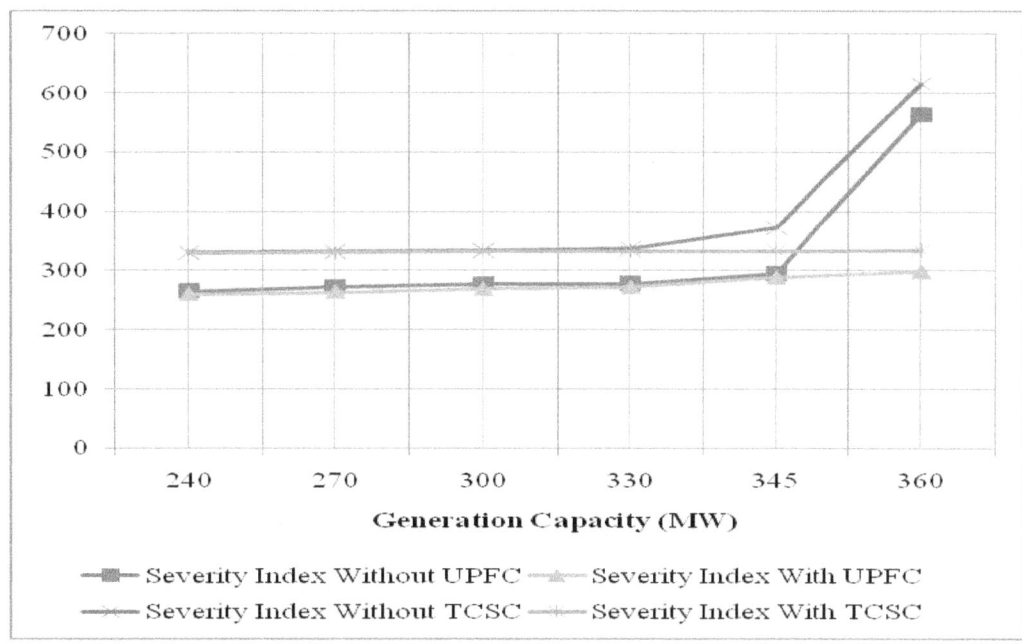

Fig 6.8: System Indices (BPECI) vs Generation Capacity

6.4 Comparison with respect to Probability of Failure & EENS

Probability of Failure & EENS are one of the important parameters to determine the relibility analysis of the given system. Probability of Failure & EENS for 6 bus RBTS are calculated at each and every bus by applying different modules of FACTS devices which have been discussed already in Chapters 4 & 5. A comparison for Probability of Failure & EENS are determined in Table 6.7 when considering 7 modules of UPFC, TCSC at all the 6 buses & demonstarted in Figs. 6.9 & 6.10 respectively. Similarly, A comparison for Probability of Failure & EENS are also detremined in Tables 6.8, 6.9, 6.10, 6.11 & 6.12 respectively when using different modules of UPFC, TCSC at all the 6 buses in the given system & demonstrated in Figs. 6.11, 6.12, 6.13, 6.14, 6.15, 6.16, 6.17, 6.18, 6.19, 6.20, 6.21 & 6.22 respectively.

Table 6.7: Comparison of Probability of Failure & EENS vs Bus Number of 6 bus RBTS

Bus No.	Probability of Failure		EENS	
	UPFC	TCSC	UPFC	TCSC
1	0.0081547	0.008411	124.64	129.94
2	0.0082665	0.008463	88.082	99.98
3	0.0083131	0.008458	377.731	395.43
4	0.0083139	0.008456	177.28	183.94
5	0.0083145	0.008457	88.86	97.58
6	0.0084512	0.008472	288.36	298.66

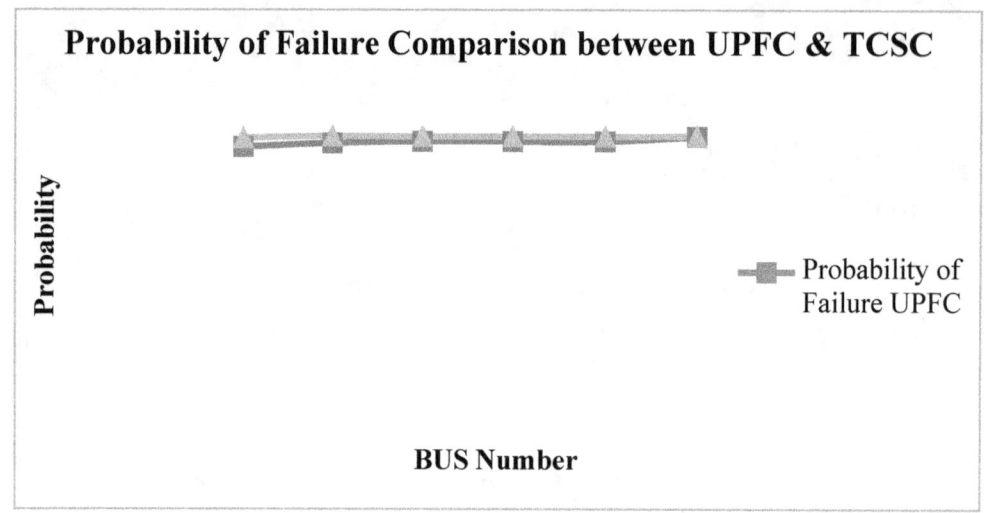

Fig 6.9: Probability of Failure Comparison between UPFC & TCSC vs Bus No. of RBTS

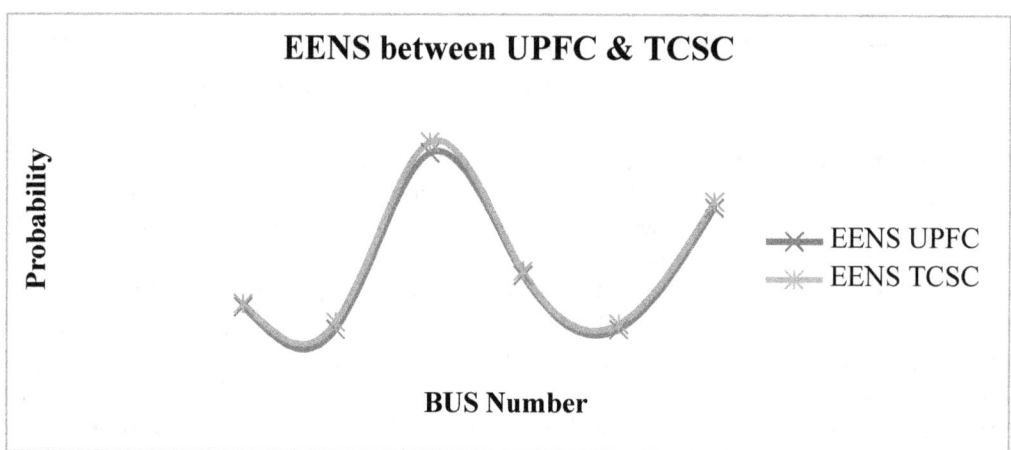

Fig 6.10: EENS Comparison between UPFC & TCSC vs Bus No. of RBTS

From Table 6.7, it can observed that Probability of Failure and EENS are more in TCSC when compared with UPFC with respect to the bus numbers of RBTS. Graphical representation of Table 6.7 is shown in Figs. 6.9 & 6.10 respectively.

Table 6.8: Comparison of Probability of Failure at Bus 1 & 2 vs No. of Modules

Module No.	Bus No. 1		Bus No. 2	
	UPFC	TCSC	UPFC	TCSC
2	0.008232	0.008536	0.008363	0.008516
3	0.008215	0.008511	0.008348	0.008507
4	0.008194	0.008459	0.008331	0.008498
5	0.008179	0.008416	0.008316	0.008491
6	0.008158	0.008411	0.008292	0.008463
7	0.008154	0.008417	0.008266	0.008475
8	0.008162	0.008426	0.008297	0.008494

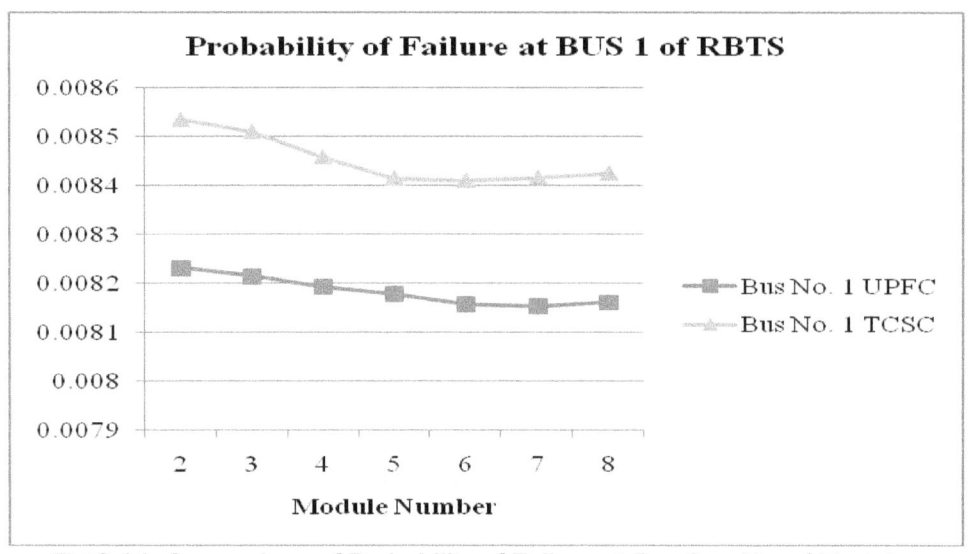

Fig 6.11: Comparison of Probability of Failure at Bus 1 vs No. of Modules

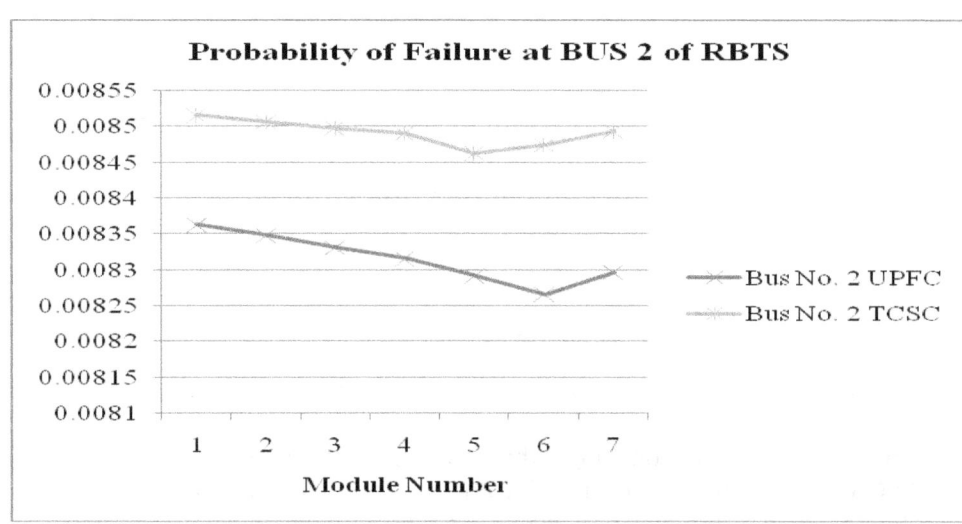

Fig 6.12: Comparison of Probability of Failure at Bus 2 vs No. of Modules

From Table 6.8, it can observed that Probability of Failure in bus number 1 and 2 is more in TCSC when compared with UPFC with respect to the number of modules. Graphical representation of Table 6.8 is shown in Figs. 6.11 & 6.12 respectively.

Table 6.9: Comparison of Probability of Failure at Bus 3 & 4 vs No. of Modules

Module No.	Bus No. 3		Bus No. 4	
	UPFC	TCSC	UPFC	TCSC
2	0.008141	0.008513	0.008137	0.008511
3	0.008167	0.008501	0.008162	0.008498
4	0.008248	0.008489	0.008245	0.008486
5	0.008281	0.008481	0.008275	0.008479
6	0.008307	0.008458	0.008306	0.008456
7	0.008313	0.008474	0.008313	0.008472
8	0.008301	0.008496	0.008302	0.008494

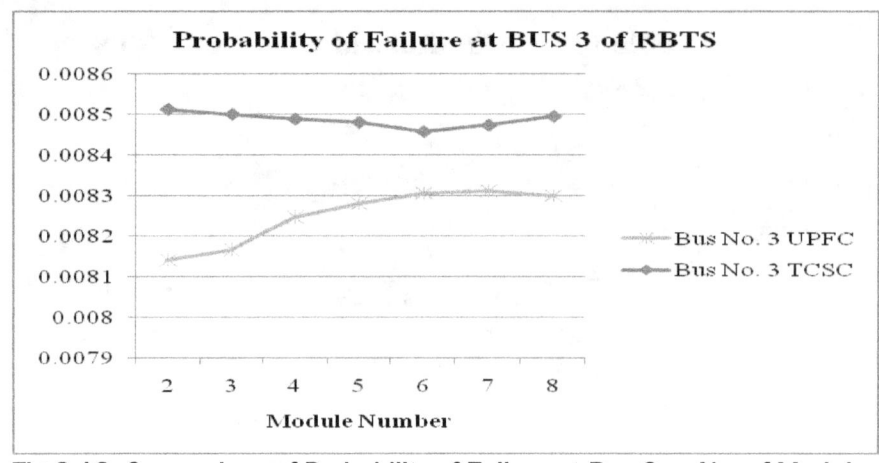

Fig 6.13: Comparison of Probability of Failure at Bus 3 vs No. of Modules

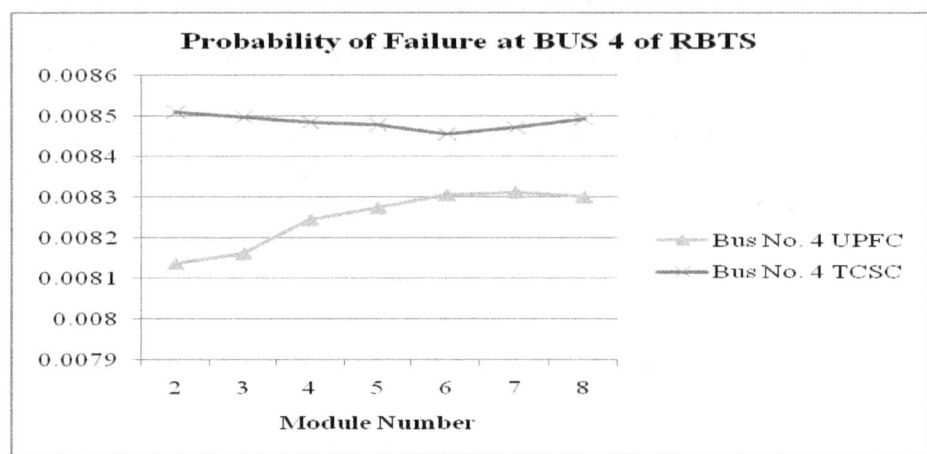

Fig 6.14: Comparison of Probability of Failure at Bus 4 vs No. of Modules

From Table 6.9, it can observed that Probability of Failure in bus number 3 and 4 is more in TCSC when compared with UPFC with respect to the number of modules. Graphical representation of Table 6.9 is shown in Figs. 6.13 & 6.14 respectively.

Table 6.10: Comparison of Probability of Failure at Bus 5 & 6 vs No. of Modules

Module No.	Bus No. 5		Bus No. 6	
	UPFC	TCSC	UPFC	TCSC
2	0.008142	0.008507	0.008634	0.00876
3	0.008171	0.008496	0.008602	0.00886
4	0.008252	0.008489	0.008562	0.00894
5	0.008281	0.008480	0.008511	0.00907
6	0.008307	0.008457	0.008472	0.00925
7	0.008314	0.008470	0.008489	0.00945
8	0.008306	0.008491	0.008496	0.00916

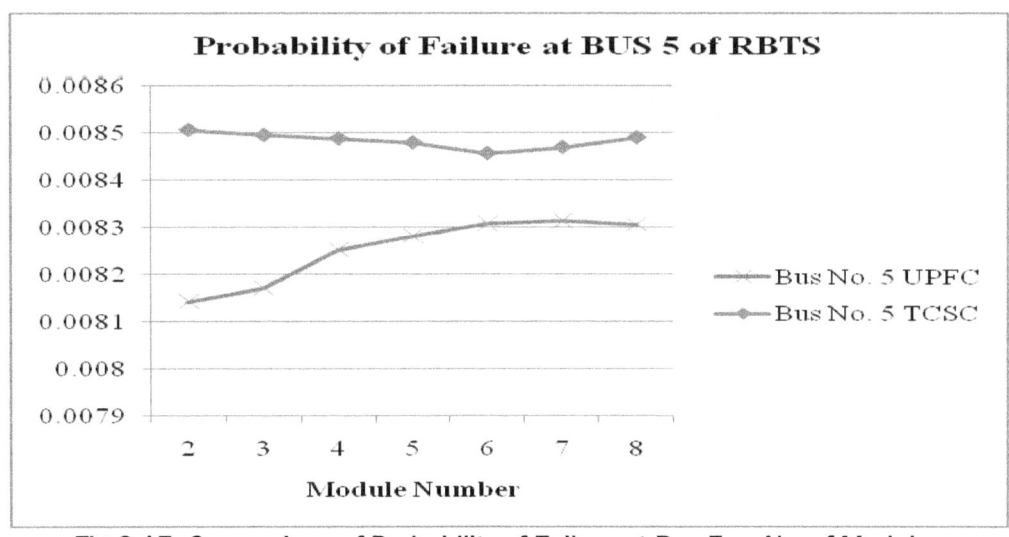

Fig 6.15: Comparison of Probability of Failure at Bus 5 vs No. of Modules

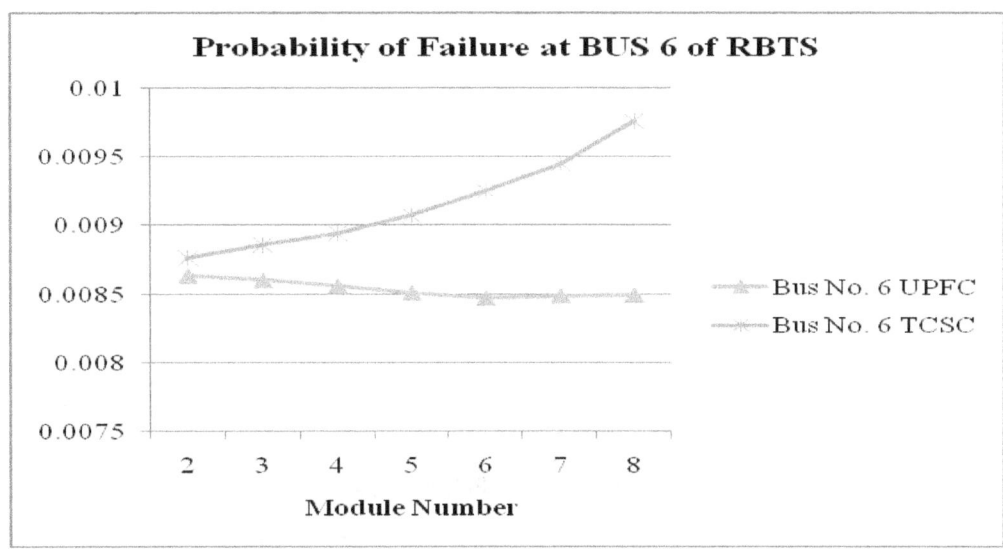

Fig 6.16: Comparison of Probability of Failure at Bus 6 vs No. of Modules

From Table 6.10, it can observed that Probability of Failure in bus number 5 and 6 is more in TCSC when compared with UPFC with respect to the number of modules. Graphical representation of Table 6.10 is shown in Figs. 6.15 & 6.16 respectively.

Table 6.11: Comparison of EENS at Bus 1, 2 & 3 vs No. of Modules

Module No.	Bus No. 1		Bus No. 2		Bus No. 3	
	UPFC	TCSC	UPFC	TCSC	UPFC	TCSC
2	126.21	132.64	89.21	102.34	377.86	398.64
3	125.94	132.12	88.81	101.96	377.35	398.14
4	125.54	131.54	88.39	101.12	376.94	397.58
5	125.12	131.08	88.12	100.78	376.61	396.91
6	124.65	129.94	87.88	99.98	376.21	395.43
7	123.97	130.16	87.68	100.36	375.92	396.16
8	124.21	130.84	87.91	101.03	376.14	396.88

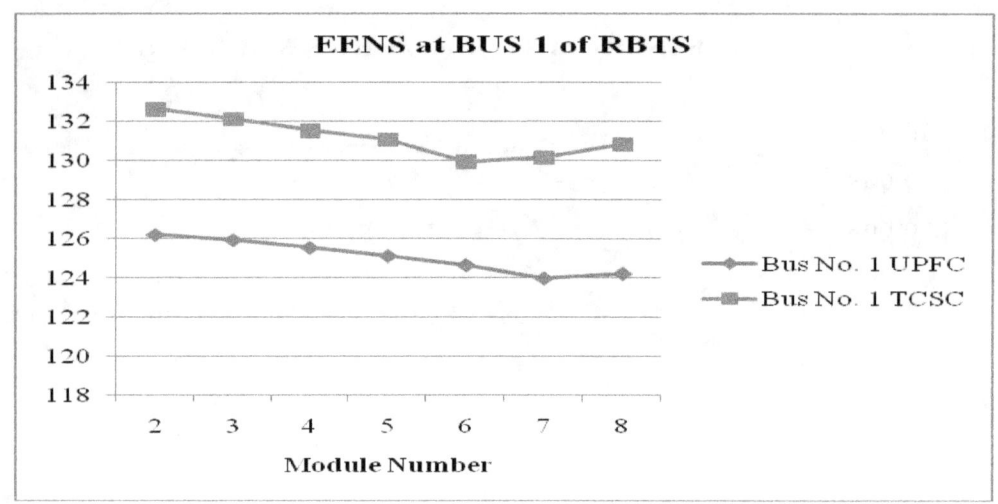

Fig 6.17: Comparison of EENS at Bus 1 vs No. of Modules

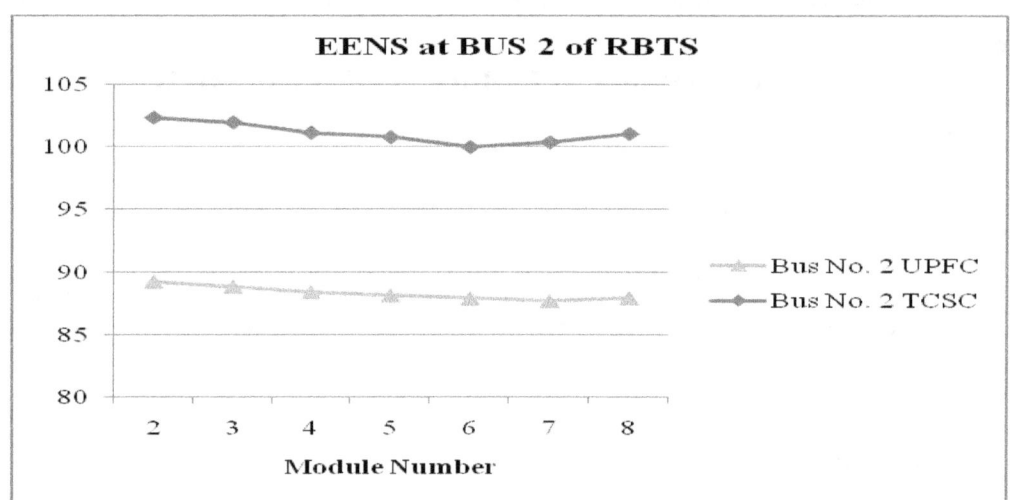

Fig 6.18: Comparison of EENS at Bus 2 vs No. of Modules

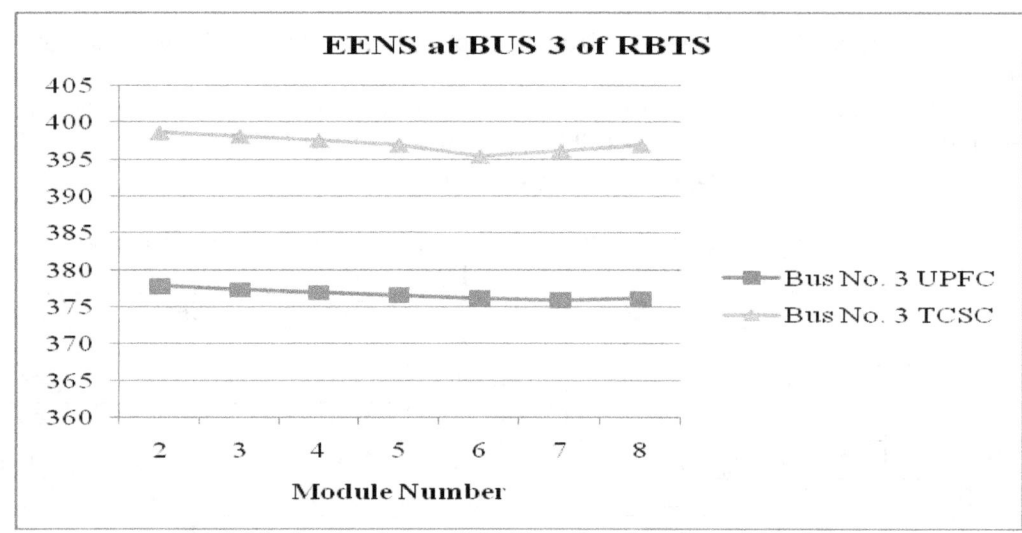

Fig 6.19: Comparison of EENS at Bus 3 vs No. of Modules

From Table 6.11, it can observed that EENS in bus number 1, 2 and 3 is more in TCSC when compared with UPFC with respect to the number of modules. Graphical representation of Table 6.10 is shown in Figs. 6.17, 6.18 & 6.19 respectively.

Table 6.12: Comparison of EENS at Bus 4, 5 & 6 vs No. of Modules

Module No.	Bus No. 4		Bus No. 5		Bus No. 6	
	UPFC	TCSC	UPFC	TCSC	UPFC	TCSC
2	178.38	186.42	90.68	100.24	290.34	302.84
3	177.97	185.96	90.31	99.38	289.97	301.63
4	177.69	185.13	89.81	98.66	289.34	300.75
5	177.26	184.66	89.42	97.81	288.94	299.54
6	176.94	183.94	88.93	97.58	288.56	298.66
7	176.54	184.52	88.49	98.12	287.99	299.18
8	176.85	185.66	88.81	99.42	288.31	300.11

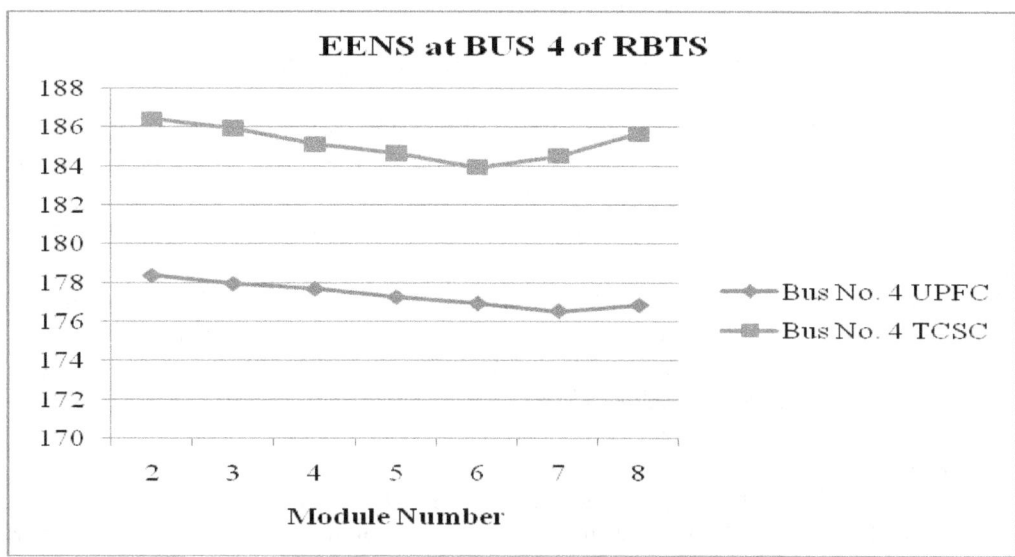

Fig 6.20: Comparison of EENS at Bus 4 vs No. of Modules

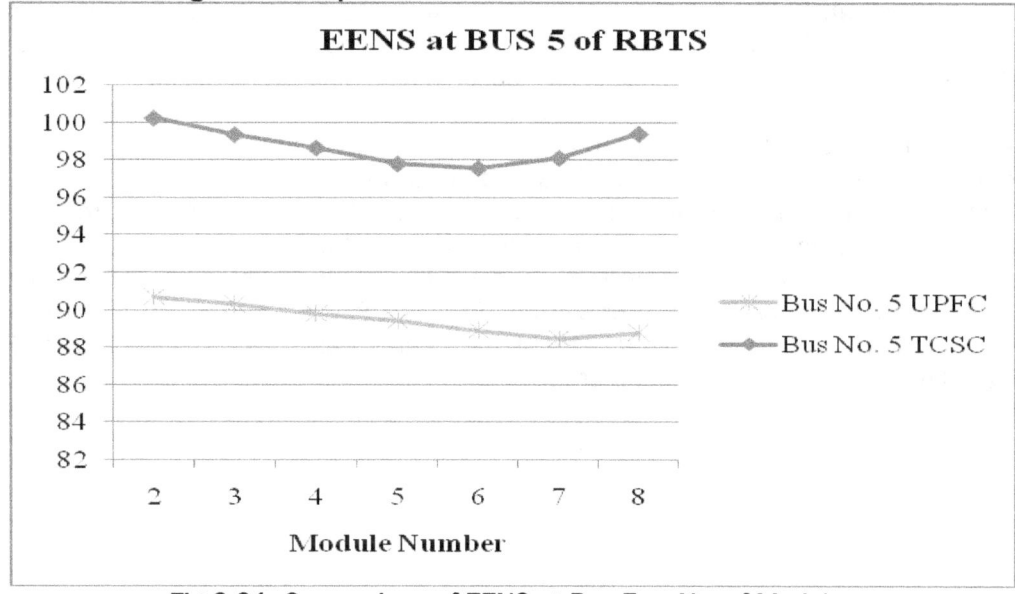

Fig 6.21: Comparison of EENS at Bus 5 vs No. of Modules

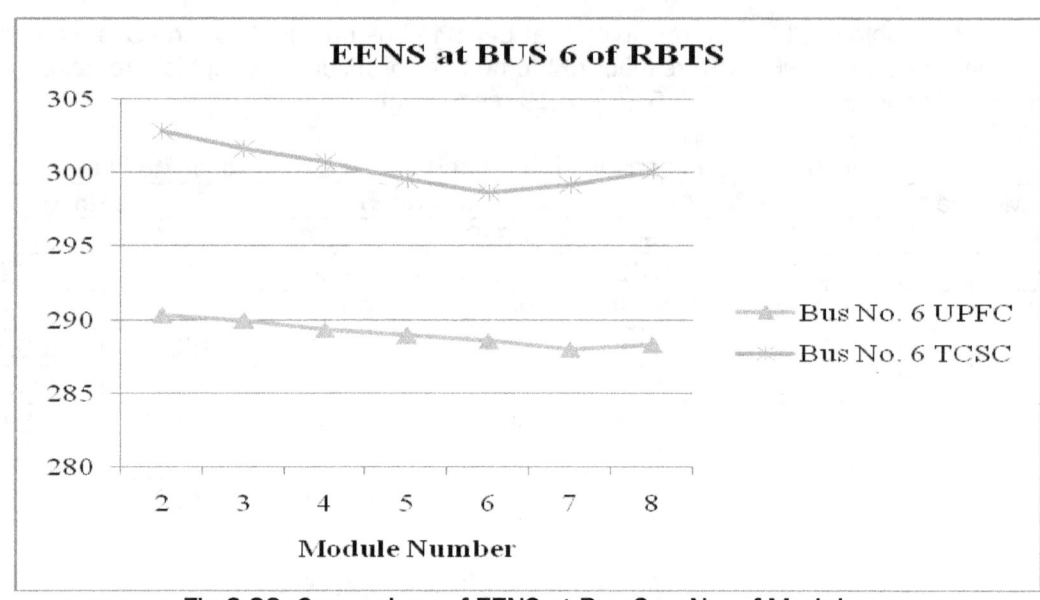

Fig 6.22: Comparison of EENS at Bus 6 vs No. of Modules

From Table 6.12, it can observed that EENS in bus number 4, 5 and 6 is more in TCSC when compared with UPFC with respect to the number of modules. Graphical representation of Table 6.12 is shown in Figs. 6.20, 6.21 & 6.22 respectively.

From Table 6.11 & 6.12, a comparison at each bus is described when TCSC & UPFC are used by changing the number of modules in the parameter of EENS. From the tables it is clear that the expected energy not supplied to the load is less when UPFC is incorporated in the composite system rather than using TCSC. The percentage change of EENS from TCSC to UPFC is 20% on an average at all the buses. Hence it can be stated that 20% of the energy has been recovered by UPFC when compared with TCSC.

6.5 Conclusion

In this chapter, comparison has been made between TCSC & UPFC for 6 bus RBTS in different aspects like, availability, system indices, probability of failure, EENS etc. From the above results, major improvement can be observed in the system when using different modules of UPFC rather than TCSC at all the buses in the system. As the components of practical model UPFC is less than TCSC the availability of UPFC is greater than TCSC, similarly the unavailability of TCSC is greater than UPFC which is clearly observed in Table 6.1. From Table 6.2, 6.3 & 6.4 it can be observed that BPSD, BPII and BPECI are less in the system when UPFC is incorporated rather than TCSC for different modules respectively. From Table 6.5 & 6.6 it is clear that, the system indices like BPII, BPSD and BPECI are reducing when UPFC & TCSC is incorporated into the system as compared with the same system without UPFC & TCSC for different Generation capacity & Load demands. Similarly, from Table 6.11 & 6.12, a comparison at each bus is described when TCSC & UPFC are used by changing the number of modules in the parameter of EENS. The percentage change of EENS from TCSC to UPFC is 20% on an average at all the buses. Hence it can be stated that 20% of the energy has been recovered by UPFC when compared with TCSC in the parameter of EENS.

Chapter 7:

RELIABILITY ANALYSIS OF COMPOSITE POWER SYSTEM USING TCSC & UPFC

7.1 Introduction

The reliability analysis of 6 Bus RBTS has been determined by using different FACTS elements like Thyristor Controlled Series Compensator (TCSC) and Unified Power Flow Controller (UPFC) independently. It can be observed from the earlier discussion that, as the number of modules used in the system are more in number, which decreases the availability of the system for transferring the same amount of the power. Since the number of components is more in number, losses in the system like switching, heat losses etc. also increase making the efficiency of the system to decrease.

In order to overcome the above criteria, it is proposed to implement both TCSC & UPFC in a single combination and incorporate in the system. Power transfer capability, compensation of the system doesn't changes as the FACTS devices have their own characteristics. In view of, the total power generated & the load in the system it is proposed to have a combination of 3 modules TCSC & 3 modules UPFC (or) 3 modules TCSC & 4 modules UPFC for additional power transfer. Here, 3 modules TCSC & 3 modules UPFC is defined as Stage 1 and 3 modules TCSC & 4 modules UPFC is defined as Stage 2. Availability & Unavailability of the two stages can be determined by using State space representation technique.

Stage 1: 3 Modules TCSC – 3 * 40MW – 120MW
3 Modules UPFC – 3 * 40MW – 120MW
Total – 240MW

Stage 2: 3 Modules TCSC – 3 * 40MW – 120MW
4 Modules UPFC – 4 * 40MW – 160MW
Total – 280MW

In this Chapter, the reliability analysis of Composite Power System using TCSC & UPFC is presented for IEEE 6 Bus RBTS and IEEE 24 Bus RTS. The reliability analysis for the 6 Bus RBTS is discussed by series-parallel representation using network reduction techniques in Section 7.2. However, network reduction techniques cannot be applied for all the systems where the availability of the system should be predicted accurately, state space representation will be used in place of network reduction techniques. The State Space representation of the combination, illustrations is discussed in Section 7.3. In Section 7.3, RLD for Stage 1 and Stage 2 are also discussed along with their results. System Indices BPSD, BPII and BPECI of 6 Bus RBTS is discussed in Section 7.4. Probability of Failure and EENS of 6 Bus RBTS is discussed in Section 7.5 along with the results. Reliability analysis of IEEE 24 Bus RTS is determined by State space representation of Stage 1 and Stage 2 using Reliability Logic Diagram, System Indices viz. BPSD, BPII & BPECI, Probability of Failure and EENS are calculated and the results are presented in Section 7.6. Conclusions of Chapter 7 are presented in Section 7.7.

7.2 Reliability Logic Diagram using Series – Parallel System

In Fig. 7.1, the Reliability Logic Diagram (RLD) of Thyristor Controlled Series Compensator and Unified Power flow Controller using Series – Parallel system is shown. Each rectangle block in the figure represents a particular component. Here each component has its own reliability which is independent of the time. Considering these reliabilities, in combination of simple series and parallel system, the overall reliability and unreliability of the system are determined as follows:

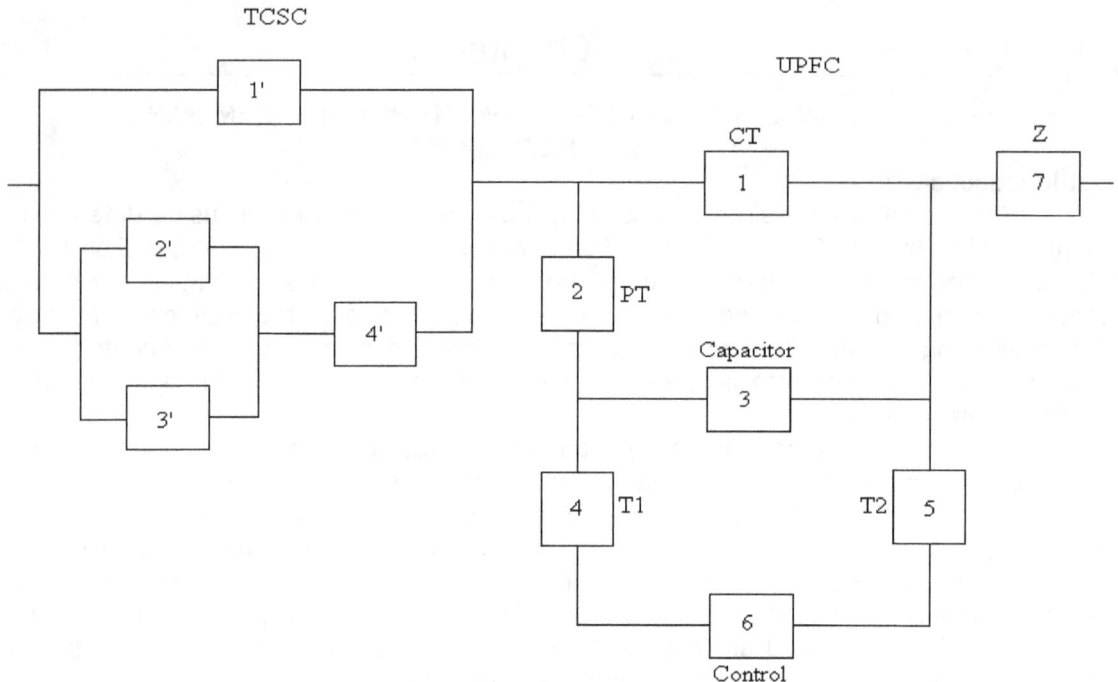

Fig. 7.1: RLD for combination of TCSC & UPFC using Series-Parallel System

7.2.1 Results
Considering individual reliabilities of each component, as given in Section 2.2.1, and the remaining components data is as follows:

    Varistor = 0.96   Reactor = 0.88
    Bypass CB = 0.84   Earth Fault CT = 0.82
    Capacitor (TCSC) = 0.85   Thyristor = 0.78

Substituting all the reliability values in the Eqns. (7.1 to 7.3)
    Reliability R = **0.968616**
    Unreliability Q = **0.031383**

7.3 Reliability Logic diagram using State Space
The State Space representation for stage 1 of the combination of TCSC and UPFC is shown in Fig. 7.2. This is another method for finding the reliability of entire system. Here the blocks from 1 to 9 represent the transition state with combination of TCSC & UPFC. The upper transition represents the states of UPFC and lower transitions represent states of TCSC. Here, only 9 states are considered for the combination of both elements and the remaining is not considered because, the remaining states cannot with stand the rated capacity of the transmission line.

The reliability logic diagram consists of two spares one of TCSC and the other of UPFC, because, each state is a combination of these two elements. Each and every state is connected to bypass module, because, at any state, the system can be failed due to any faults or improper firing of thyristors or failure of capacitors in series compensator. Each state is assigned with a proper transition number in a sequential manner so that, each state follows the previous states. A bypass block is also considered, to reduce the capacitor voltage which is due to fault currents.

7.3.1 Stage 1

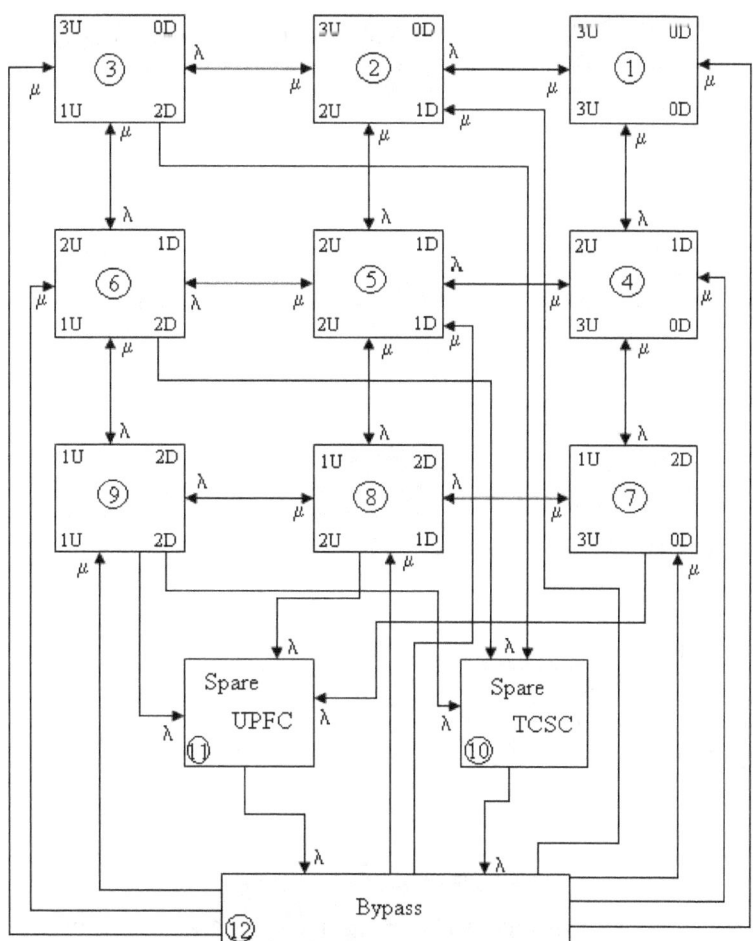

Fig. 7.2: RLD for Combination of TCSC & UPFC (Stage 1) using State – Space Representation for 6 bus RBTS

From the above reliability logic diagram, define matrix P, which is Stochastic Transitional Probability Matrix (STPM).

$$P = \begin{bmatrix}
1-2\lambda & \lambda & 0 & \lambda & 0 & 0 & 0 & 0 & 0 & 0 & 0 & 0 \\
\mu & 1-(3\lambda+\mu) & \lambda & 0 & \lambda & 0 & 0 & 0 & 0 & \lambda & 0 & 0 \\
0 & \mu & 1-(2\lambda+\mu) & 0 & 0 & \lambda & 0 & 0 & 0 & \lambda & 0 & 0 \\
\mu & 0 & 0 & 1-(3\lambda+\mu) & \lambda & 0 & \lambda & 0 & 0 & \lambda & 0 \\
0 & \mu & 0 & \mu & 1-(4\lambda+2\mu) & \lambda & 0 & \lambda & 0 & \lambda & \lambda & 0 \\
0 & 0 & \mu & 0 & \mu & 1-(3\lambda+2\mu) & 0 & 0 & \lambda & \lambda & \lambda & 0 \\
0 & 0 & 0 & \mu & 0 & 0 & 1-(2\lambda+\mu) & \lambda & 0 & 0 & \lambda & 0 \\
0 & 0 & 0 & 0 & \mu & 0 & \mu & 1-(3\lambda+2\mu) & \lambda & \lambda & \lambda & 0 \\
0 & 0 & 0 & 0 & 0 & \mu & 0 & \mu & 1-(2\lambda+\mu) & \lambda & \lambda & 0 \\
0 & 0 & 0 & 0 & 0 & 0 & 0 & 0 & 0 & 1-\lambda & 0 & \lambda \\
0 & 0 & 0 & 0 & 0 & 0 & 0 & 0 & 0 & 0 & 1-\lambda & \lambda \\
\mu & \mu & \mu & \mu & \mu & \mu & \mu & \mu & \mu & 0 & 0 & 1-9\mu
\end{bmatrix}$$

[P$_{SS}$] [P] = [P$_{SS}$]

Where P$_{SS}$ = [P$_1$ P$_2$ .... P$_{11}$ P$_{12}$] which is a limiting state probability vector. Expressing the above matrix form in terms of equations:

$$P_1(1-2\lambda) + P_2\mu + P_4\mu + P_{12}\mu = P_1 \quad \quad (7.1)$$

$$P_1\lambda + P_2(1-(\mu+3\lambda)) + P_5\mu + P_{12}\mu = P_2 \quad \quad (7.2)$$

$$P_2\lambda + P_3(1-(\mu+2\lambda)) + P_6\mu + P_{12}\mu = P_3 \quad \quad (7.3)$$

$$P_1\lambda + P_4(1-(\mu+3\lambda)) + P_5\mu + P_7\mu + P_{12}\mu = P_4 \quad \quad (7.4)$$

$$P_2\lambda + P_4\lambda + P_5(1-(2\mu+4\lambda)) + P_6\mu + P_8\mu + P_{12}\mu = P_5 \quad \quad (7.5)$$

$$P_3\lambda + P_5\lambda + P_6(1-(2\mu+3\lambda)) + P_9\mu + P_{12}\mu = P_6 \quad \quad (7.6)$$

$$P_4\lambda + P_7(1-(2\mu+4\lambda)) + P_8\mu + P_{12}\mu = P_7 \quad \quad (7.7)$$

$$P_5\lambda + P_7\lambda + P_8(1-(2\mu+3\lambda)) + P_9\mu + P_{12}\mu = P_8 \quad \quad (7.8)$$

$$P_6\lambda + P_8\lambda + P_9(1-(2\mu+2\lambda)) + P_{12}\mu = P_9 \quad \quad (7.9)$$

$$P_2\lambda + P_3\lambda + P_5\lambda + P_6\lambda + P_8\lambda + P_9\lambda + P_{10}(1-\lambda) = P_{10} \quad \quad (7.10)$$

$$P_4\lambda + P_5\lambda + P_6\lambda + P_7\lambda + P_8\lambda + P_9\lambda + P_{11}(1-\lambda) = P_{11} \quad \quad (7.11)$$

$$P_{10}\lambda + P_{11}\lambda + P_{12}(1-9\mu) = P_{12} \quad \quad (7.12)$$

Since all the above Eqns. (7.1 to 7.12) are independent to each other, we consider only 11 equations out of the above 12 equations and 12th equation is to be taken as P$_1$+P$_2$.....................+P$_{11}$+P$_{12}$ = 1 . . . (7.13)

7.3.1.1 Results

From the above, the Limiting State Probabilities can be obtained.
Consider the data: Failure rate (λ) = 0.7 f/yr
Repair Rate (μ) = 150 hrs of each component, then
Individual LSPs are:
P$_1$ = 0.97642   P$_2$ = 0.012402   P$_3$ = 0.00025   P$_4$ = 1.3548*10^{-3}
P$_5$ = 2.709*10^{-4} P$_6$ = 5.4194*10^{-5} P$_7$ = 3.847*10^{-7} P$_8$ = 4.6684*10^{-8}
P$_9$ = 6.7134*10^{-9} P$_{10}$ = 0.008524   P$_{11}$ = 0.008524   P$_{12}$ = 8.3216*10^{-12}

P$_{UP}$ = P$_1$ + P$_{10}$ + P$_{11}$ = 0.97642 + 0.008524 + 0.008524 = **0.985666**
P$_{DOWN}$ = 1 − P$_{UP}$ = **0.014334**

7.3.2 Stage 2

The state space representation for stage 2 of combination of TCSC and UPFC is shown in Fig. 7.3. In this Fig. 7.3, the blocks 1 to 15 represent transition states. The upper transition rates are of UPFC and lower transitional rates are of TCSC. Here, 12 states are considered because, the remaining states will represent the failed states as they cannot withstand rated capacity.

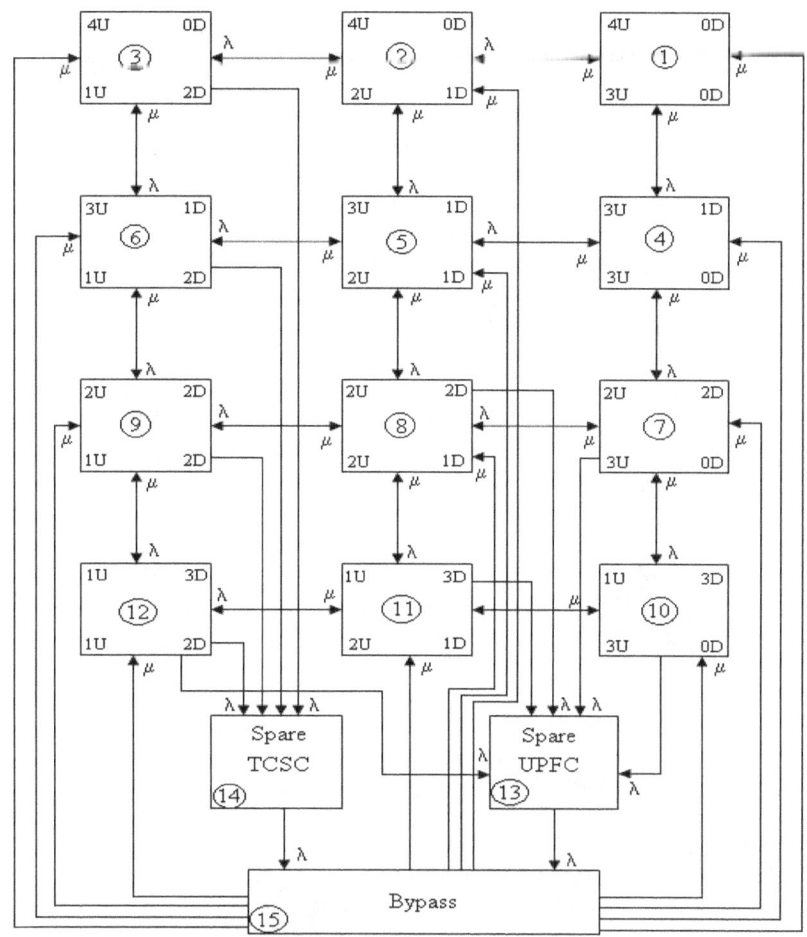

Fig. 7.3: RLD for Combination of TCSC & UPFC (Stage 2) using State – Space Representation for 6 bus RBTS

7.3.2.1 Results

From the Eqns. (7.1 to 7.11) & Eqn. (7.16) find the Limiting State Probabilities
Consider the data: Failure rate (λ) = 0.7 f/yr
      Repair Rate (μ) = 150 hrs of each component, then
Individual LSPs are:

P_1 = 0.96564 P_2 = 0.0231642 P_3 = 0.000396 P_4 = 1.3548*10^{-3}
P_5 = 2.709*10^{-4} P_6 = 5.4194*10^{-5} P_7 = 3.847*10^{-6} P_8 = 4.6684*10^{-7}
P_9 = 6.7134*10^{-8} P_{10} = 5.6879*10^{-9} P_{11} = 9.64786*10^{-10} P_{12} = 8.3216*10^{-12}
P_{13} = 0.003946 P_{14} = 0.003946 P_{15} = 9.1326*10^{-14}
 P_{UP} = P_1 + P_{10} + P_{11} = 0.96564 + 0.003946 + 0.003946 = **0.973532**
 P_{DOWN} = 1 – P_{UP} = **0.026468**
In Table 7.1, the results for stage 1 & stage 2 are presented.

Table 7.1: Availability & Unavailability of different Stages

Stage	Modules		Availability	Unavailability
	TCSC	UPFC		
1	3	3	0.985666	0.014334
2	3	4	0.973532	0.0264628

From Table 7.1, it can be observed that as the no. of stages increase, the availability will decrease although it satisfies the required performance.

7.4 System Indices

System Indices like BPSD, BPII & BPECI are calculated for the 6 bus RBTS system by incorporating the combination of FACTS devices as shown in Table 7.2

For Stage 1: (from Fig. 7.2)

Bulk Power Supply average curtailment / disturbance, using Eqn. (4.12) is obtained as

$$= \frac{82.48 * 0.9934}{5.562} = 14.73 \text{ MW/disturbance}$$

Bulk Power Interruption Index, using Eqn. (4.13) is obtained as

$$= \frac{82.48 * 0.9934}{240} = 0.3414 \text{ MW / MW-yr}$$

Bulk Power Energy Curtailment index (Severity Index), using Eqn. (4.14) is obtained as

$$= 60 * \frac{43.9 * 0.99245 * 20.76}{240} = 229.66 \text{ MWh/MW-yr}$$

For Stage 2: (from Fig. 7.3)

Bulk Power Supply average MW curtailment / disturbance

$$= \frac{82.039 * 0.99289}{5.5525} = 14.67 \text{ MW/disturbance}$$

Bulk Power Interruption Index $= \frac{82.039 * 0.99289}{240} = 0.3394 \text{ MW / MW-yr}$

Bulk Power Energy Curtailment index (Severity Index)

$$= 60 * \frac{44.224 * 0.99225 * 20.64}{240} = 226.43 \text{ MWh/MW-yr}$$

In Table 7.2, the system indices for different stages are presented from Figs. 7.2 & 7.3. From the Table 7.2, it can be observed that as the no. of stages increase, the system indices are decreasing, i.e., performance of the system is increasing.

Table 7.2: System Indices of 6 bus RBTS with different Stages

Stage	BPSD	BPII	BPECI
1	14.73	0.3414	229.66
2	14.67	0.3394	226.43

7.5 Probability of Failure & EENS

Further, system indices, probability of failure & EENS of the system are also calculated at each bus which is presented in Table 7.3 & 7.4 and graphically in Figs. 7.4 & 7.5 respectively.

Table 7.3: Probability of Failure for 6 bus RBTS at different bus vs different Stages

Stage	Bus No.					
	1	2	3	4	5	6
1	0.0069124	0.0069011	0.0068844	0.0069645	0.0067452	0.0067312
2	0.0069087	0.0068871	0.0068814	0.0069622	0.0067184	0.0067024

From Table 7.3, it can be observed that at each bus the probablility of failure is reducing when stage 2 is incorporated rather than stage 1. If probability of the failure is decreasing, the availiability of the system increases which shows improvement in the performance of the system.

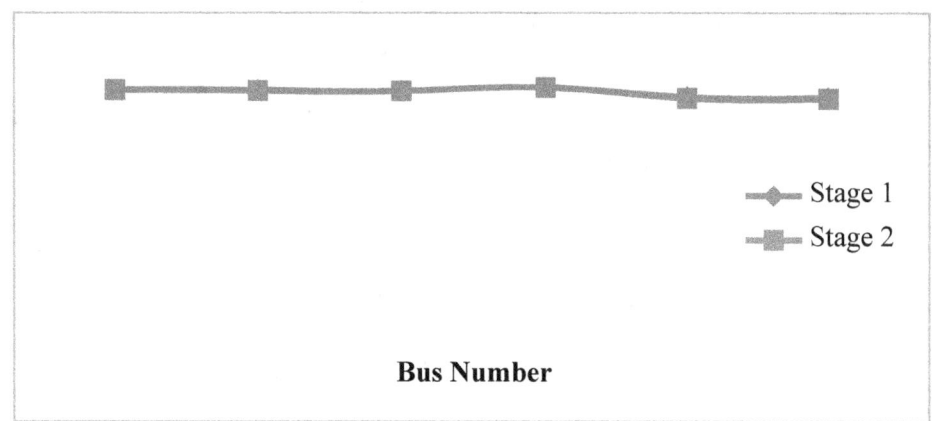

Fig. 7.4: Probability of Failure for 6 bus RBTS at different bus vs different Stages

Table 7.4: EENS for 6 bus RBTS at different bus vs different Stages

Stage	Bus No.					
	1	2	3	4	5	6
1	117.64	85.63	351.66	171.22	84.12	262.18
2	111.23	78.12	338.67	164.34	80.64	251.38

From Table 7.4, it can be observed that at each bus the Expected Energy Not Supplied is decreasing when stage 2 is incorporated rather than stage 1. The load at each bus is efficienlty used by stage 2 rather than stage 1.

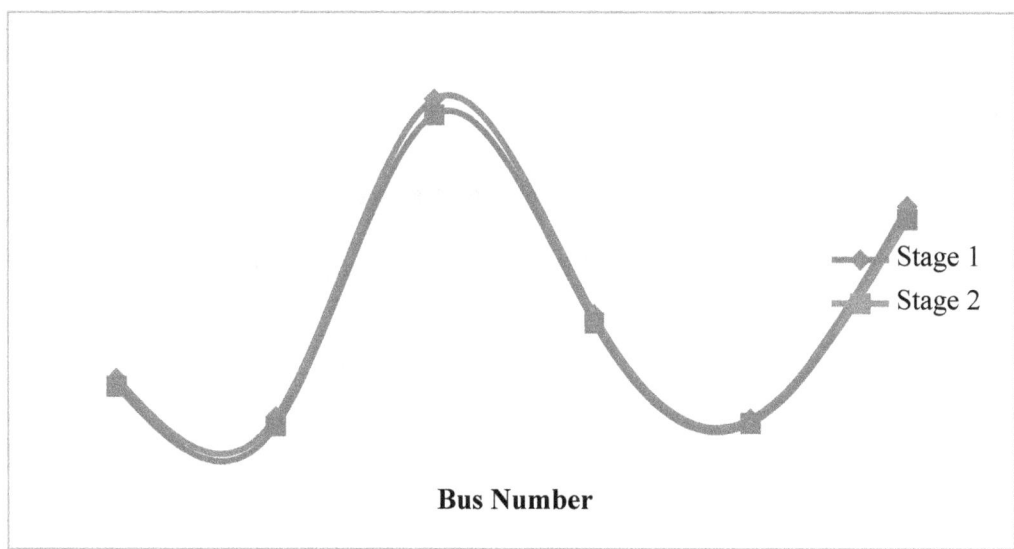

Fig. 7.5: EENS for 6 bus RBTS at different bus vs different Stages

Similarly the system indices are calculated with respect to Generation Capacity for both the stages which are tabulated in Table 7.5 & 7.6 and graphically represented in Fig. 7.6, 7.7 & 7.8 respectively.

Table 7.5: System Indices vs Generation Capacity – Stage 1

Generation Capacity (MW)	Load Demand (MW)	BPSD	BPII	BPECI
240	185	14.73	0.3414	229.66
270	203.5	14.91	0.3424	230.44
300	222	15.14	0.3457	232.45
330	240.5	15.29	0.3491	237.51
345	259	16.43	0.3534	243.12
360	277.5	18.24	0.3724	253.66

Table 7.6: System Indices vs Generation Capacity – Stage 2

Generation Capacity (MW)	Load Demand (MW)	BPSD	BPII	BPECI
240	185	14.67	0.3394	226.43
270	203.5	14.76	0.3398	228.36
300	222	15.04	0.3415	229.45
330	240.5	15.16	0.3423	230.67
345	259	16.17	0.3446	235.41
360	277.5	17.92	0.3647	240.12

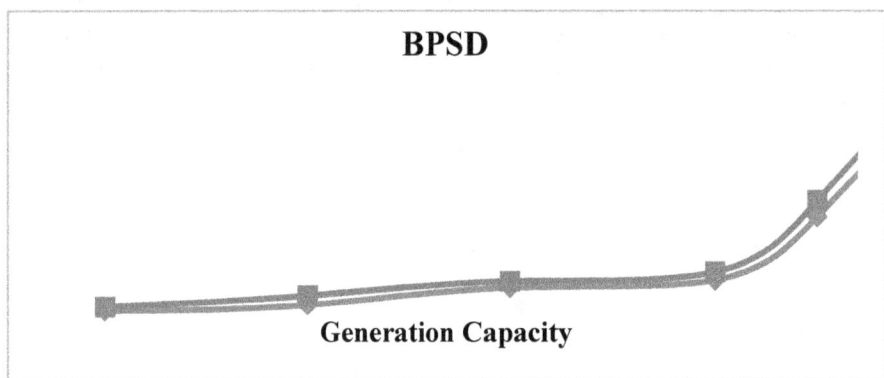

Fig. 7.6: System Indices (BPSD) at different Stages vs Generation Capacity

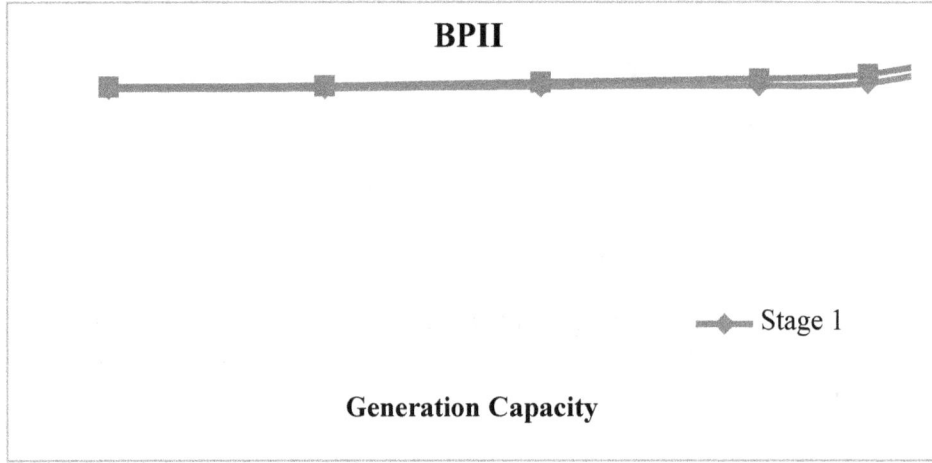

Fig. 7.7: System Indices (BPII) at different Stages vs Generation Capacity

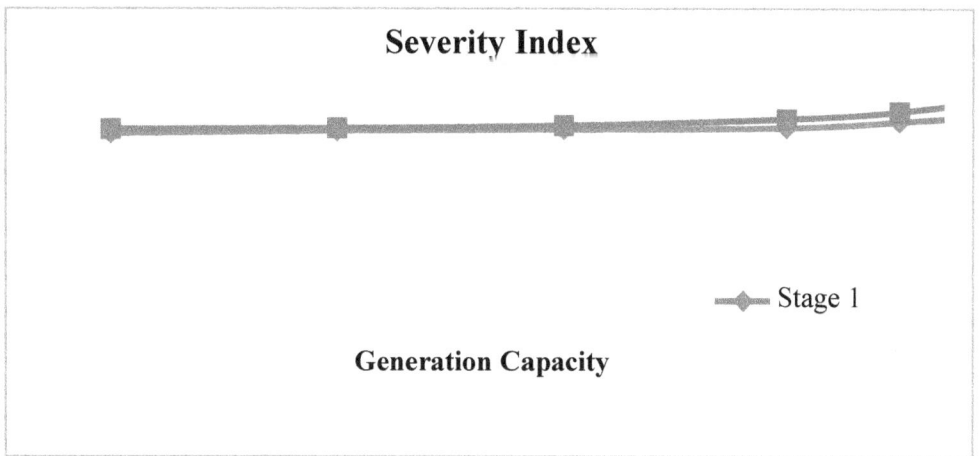

Fig. 7.8: System Indices (BPECI) at different Stages vs Generation Capacity

From Tables 7.5 and 7.6, it can be observed that the system indices BPII, BPSD & BPECI are decreasing when stage 2 is incorporated in the system. When power interruption, supply disturbance and curtailment index are reduced, the availability of the system increases which indicates healthier power system.

So far, the analysis of the sample IEEE 6 Bus RBTS is discussed. In order to generalize the inferences obtained from the sample systems, IEEE 24 Bus RTS is considered and the results are presented, with the same data as discussed in previous chapters. The results of IEEE 24 Bus system are presented in Section 7.6.

7.6 IEEE 24 BUS RTS:

The Single diagram of IEEE 24 Bus Reliability Test System (RTS) is shown in Fig. 7.9. The line and probability of the system is shown in Appendix A. 1.4 and A. 1.5 respectively.

Average load at the buses is 235.87MW. Depending on the with stand capacity, repair rate and failure rate, it is feasible to have the combination of 3 module UPFC & 3 module TCSC in all the transmission line except in 1 to 2, 1 to 4, 1 to 5, 2 to 2, 2 to 4, 2 to 5, 3 to 2, 3 to 4 and 3 to 5 lines. In these lines the maximum power transmitted is only 97 MW [38] throughout the year. Based on the above criteria, only 1 module TCSC & 1 module UPFC are incorporated in the above 9 transmission line with 20 % increase in their individual capacities.

Stage 1: 3 Modules TCSC – 3 * 40MW – 120MW
3 Modules UPFC – 3 * 40MW – 120MW
Total – 240MW

Stage 2: with 20% increase in the individual capacity of TCSC & UPFC (for 1 to 2, 1 to 4, 1 to 5, 2 to 2, 2 to 4, 2 to 5, 3 to 2, 3 to 4 and 3 to 5 transmission lines only)
1 Module TCSC – 1 * 48MW – 48MW
1 Module UPFC – 1 * 48MW – 48MW
Total – 96MW

Stage 1 and Stage 2 are incorporated in the 24 Bus System independent of the load demand. The reliability analysis is carried out by incorporating Stage 1 and 2 simultaneously in the system. Availability and unavailability of the two stages are calculated by State Space representation

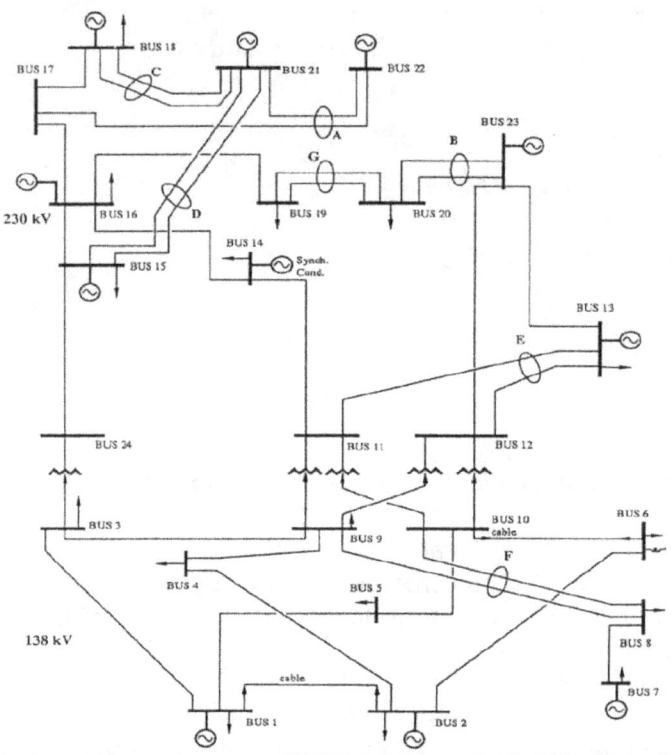

Fig 7.9: Single Line diagram of IEEE 24 Bus – Reliability Test System

7.6.1 Reliability Logic diagram using State Space

7.6.1.1 Stage 1

The Reliability Logic Diagram of IEEE 24 Bus RTS for the combination of TCSC & UPFC with 3 modules each using state space representation is shown in Fig. 7.10.

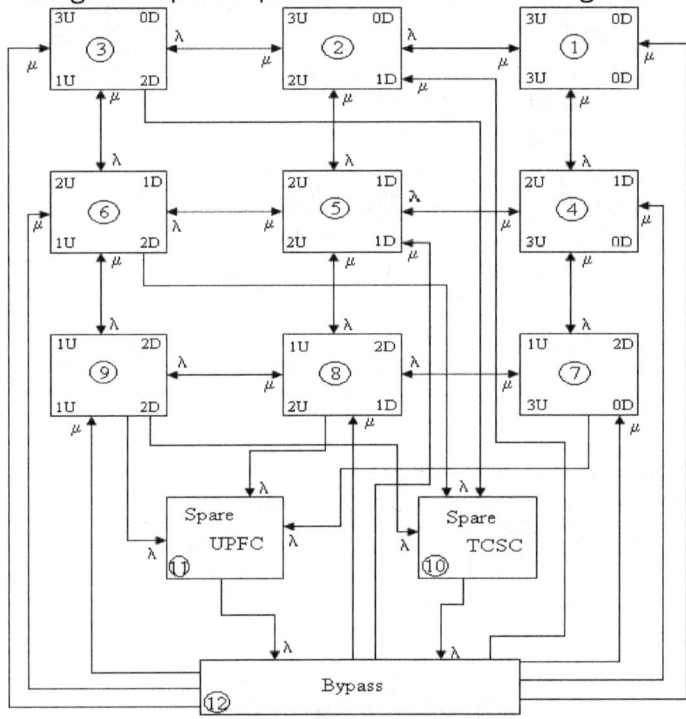

Fig. 7.10: RLD for Combination of TCSC & UPFC (Stage 1) using State – Space Representation for IEEE 24 bus RTS

7.6.1.1.1 Results
From the above, the limiting State Probabilities can be obtained.
Consider the data: Failure rate (λ) = 0.7 f/yr
Repair Rate (μ) = 150 hrs of each component, then

Individual LSPs are:
P_1 = 0.97642 $\quad\quad\quad$ P_2 = 0.012402 $\quad\quad\quad$ P_3 = 0.00025 $\quad\quad\quad$ P_4 = 1.3548*10^{-3}
P_5 = 2.709*10^{-4} $\quad$ P_6 = 5.4194*10^{-5} $\quad$ P_7 = 3.847*10^{-7} $\quad$ P_8 = 4.6684*10^{-8}
P_9 = 6.7134*10^{-9} $\quad$ P_{10} = 0.008524 $\quad\quad$ P_{11} = 0.008524 $\quad\quad$ P_{12} = 8.3216*10^{-12}
$\quad P_{UP}$ = P_1 + P_{10} + P_{11} = 0.97642 + 0.008524 + 0.008524 = **0.985666**
$\quad P_{DOWN}$ = 1 − P_{UP} = **0.014334**

The RLD using State Space representation for stage 1 is already discussed in 7.3.1. The results of stage 1 are presented

7.6.1.2 Stage 2
The state space representation for stage 2 of combination of TCSC and UPFC is shown in Fig. 7.11. In Fig. 7.11, the blocks 1 to 7 represent transition states. The upper transition rates are of UPFC and lower transitional rates are of TCSC. Here, 4 states are considered because, as the remaining states will represent the failed states as they cannot withstand rated capacity.

Stage 2: (for 1 to 2, 1 to 4, 1 to 5, 2 to 2, 2 to 4, 2 to 5, 3 to 2, 3 to 4 and 3 to 5 transmission lines only)
$\quad\quad\quad\quad$ 1 Module TCSC − 1 * 48MW − 48MW
$\quad\quad\quad\quad$ 1 Module UPFC − 1 * 48MW − 48MW
$\quad\quad\quad\quad\quad\quad\quad\quad$ Total − 96MW

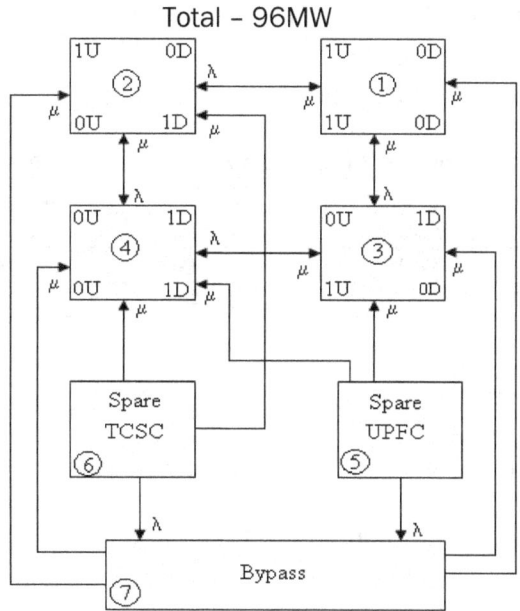

Fig. 7.11: RLD for Combination of TCSC & UPFC (Stage 2) using State − Space Representation for IEEE 24 Bus RTS

7.6.1.2.1 Results
Finding the Limiting State Probabilities
Consider the data: Failure rate (λ) = 0.7 f/yr
$\quad\quad\quad\quad\quad\quad$ Repair Rate (μ) = 150 hrs of each component, then
Individual LSPs are:
P_1 = 0.979347 $\quad\quad$ P_2 = 0.005871 $\quad\quad$ P_3 = 0.001405 $\quad\quad$ P_4 = 0.000932
P_5 = 0.005989 $\quad\quad$ P_6 = 0.005989 $\quad\quad$ P_7 = 0.000467
$\quad P_{UP}$ = P_1 + P_5 + P_6 = 0.979347 + 0.003946 + 0.003946 = **0.991325**
$\quad P_{DOWN}$ = 1 − P_{UP} = **0.008675**

In Table 7.7, the results of availability and unavailability of IEEE 24 bus RTS for stage 1 & stage 2 are presented.

Table 7.7: Availability and Unavailability of different Stages for IEEE 24 Bus RTS

Stage	Modules		Availability	Unavailability
	TCSC	UPFC		
1	3	3	0.985666	0.014334
2	1	1	0.9991325	0.008675

From Table 7.7, it can be observed that as the no. of stages increase, the availability will decrease although it satisfies the required performance.

7.6.2 System Indices

System Indices like BPSD, BPII & BPECI are calculated for IEEE 24 bus RTS system by incorporating the combination of FACTS devices.

Bulk Power Supply average curtailment / disturbance, using Eqn. (4.12) is obtained as

$$= \frac{2688.33 * 0.98912}{3.632} = 732.18 \text{ MW/disturbance}$$

Bulk Power Interruption Index, using Eqn. (4.13) is obtained as

$$= \frac{5939.609 * 0.98912}{3405} = 1.7254 \text{ MW / MW-yr}$$

Bulk Power Energy Curtailment index (Severity Index), using Eqn. (4.14) is obtained as

$$= 60 * \frac{4313.9695 * 0.986451 * 25.47}{3405} = 1909.94 \text{ MWh/MW-yr}$$

The system indices for IEEE 24 Bus RTS are presented in Table 7.8.

Table 7.8: System Indices of IEEE 24 Bus RTS with different FACTS Components

System Indices	Original	TCSC	UPFC	TCSC & UPFC
BPSD	817.22	784.56	780.19	732.128
BPII	2.620	2.0156	1.9987	1.7254
BPECI	2211.640	1987.41	1924.65	1909.94

From Table 7.8, it can be observed that the system indices viz. BPSD, BPII & BPECI are reducing when using FACTS controllers in the system (IEEE 24 bus). It can be noted that when the combination of TCSC & UPFC is incorporated in the system at different locations the system indices are gradually reduced when compared with other components.

System Indices (BPSD, BPII and BPECI) are further calculated at each bus as shown in Tables 7.9 to 7.11. The graphical forms of the Tables 7.9 to 7.11 are shown in Figs. 7.12 to 7.14.

Table 7.9: BPSD of IEEE 24 Bus at each Bus with different FACTS Components

Bus No	BPSD			
	Original	TCSC	UPFC	UPFC & TCSC
1	817.22	784.56	780.19	732.128
2	817.22	784.56	780.19	732.128
3	817.22	784.56	777.18	729.118
4	817.22	784.56	780.09	732.028
5	817.22	784.56	780.19	732.128

6	817.22	784.56	780.19	732.128
7	816.45	783.79	779.42	731.358
8	817.22	784.56	780.19	732.128
9	817.22	784.56	778.24	730.178
10	817.22	784.56	780.19	732.128
11	817.22	784.56	780.19	732.128
12	817.22	784.56	780.01	731.948
13	817.22	783.91	779.54	731.478
14	817.22	784.56	780.19	732.128
15	817.22	784.56	780.19	732.128
16	817.22	784.56	780.19	732.128
17	817.22	784.56	779.12	731.058
18	817.22	784.56	780.19	732.128
19	817.22	784.56	780.19	732.128
20	817.22	784.56	780.19	732.128
21	817.22	784.56	780.19	732.128
22	817.22	784.56	780.19	732.128
23	817.22	784.56	776.92	728.858
24	817.22	784.56	780.19	732.128

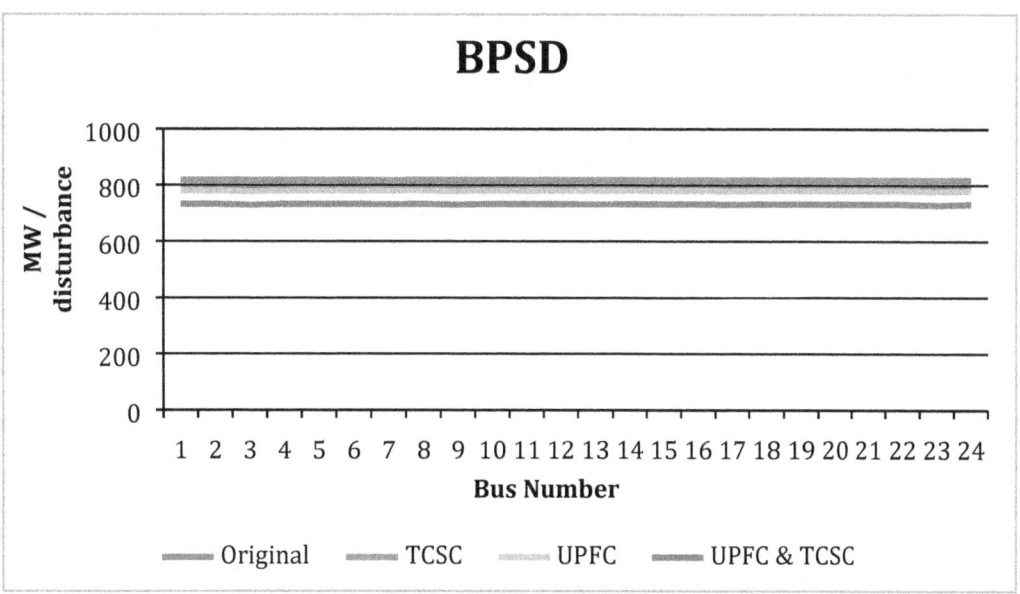

Fig. 7.12: BPSD of IEEE 24 Bus at each Bus with different FACTS Components

From Table 7.9 it can be observed that, Bulk Power Supply Disturbance is decreasing when the combination of TCSC & UPFC is incorporated into the system rather than the system when incorporated by TCSC, UPFC independently. The graphical form in Fig. 7.12 shows clearly the reduction in BPSD.

Table 7.10: BPII of IEEE 24 Bus at each Bus with different FACTS Components

Bus No	BPII			
	Original	TCSC	UPFC	UPFC & TCSC
1	2.62	2.0156	1.9987	1.7254
2	2.62	2.0156	1.9987	1.7254
3	2.62	2.0001	1.9832	1.7099
4	2.62	2.0156	1.9987	1.7254
5	2.62	2.0156	1.9987	1.7254
6	2.62	2.0156	1.9987	1.7254
7	2.611	2.0066	1.9734	1.7001
8	2.62	2.0156	1.9987	1.7254
9	2.62	2.0156	1.9987	1.7254
10	2.62	2.0156	1.9987	1.7254
11	2.62	2.0156	1.9987	1.7254
12	2.62	2.0156	1.9987	1.7254
13	2.62	2.0072	1.9903	1.717
14	2.62	2.0002	1.9833	1.71
15	2.62	2.0072	1.9903	1.717
16	2.62	2.0156	1.9987	1.7254
17	2.62	2.0156	1.9987	1.7254
18	2.62	2.0156	1.9987	1.7254
19	2.62	2.0156	1.9987	1.7254
20	2.62	2.0156	1.9987	1.7254
21	2.62	2.0156	1.9987	1.7254
22	2.62	2.0156	1.9987	1.7254
23	2.62	2.0156	1.9753	1.702
24	2.62	2.0156	1.9987	1.7254

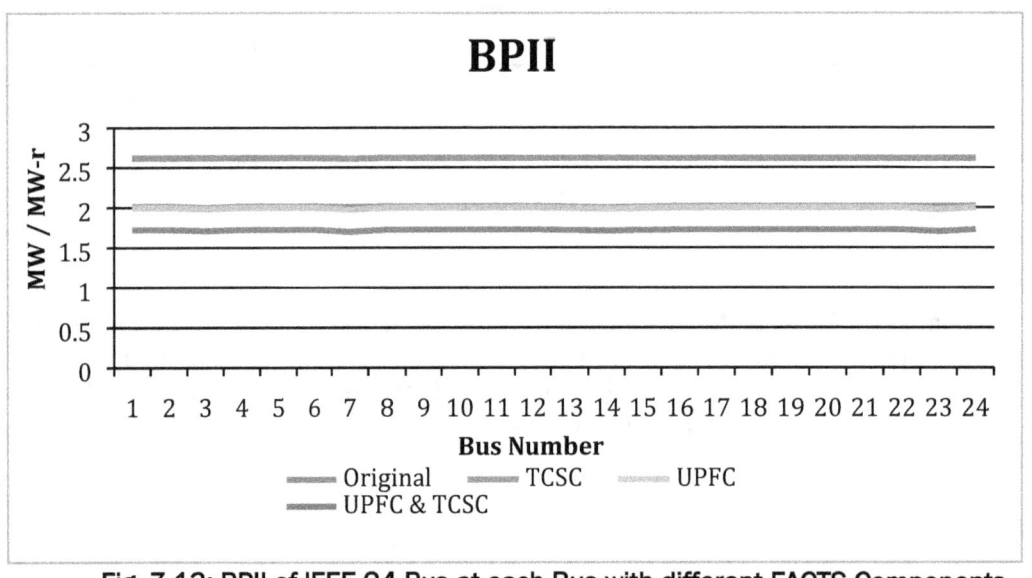

Fig. 7.13: BPII of IEEE 24 Bus at each Bus with different FACTS Components

From Table 7.10 it can be observed that, Bulk Power Interruption Index is decreasing when the combination of TCSC & UPFC is incorporated into the system rather than the system when incorporated by TCSC, UPFC independently. Once the Interruption Index is decreasing obviously the system performance increases. The graphical form in Fig. 7.13 shows clearly the reduction in BPII.

Table 7.11: BPECI of IEEE 24 Bus at each Bus with different FACTS Components

Bus No	BPECI			
	Original	TCSC	UPFC	UPFC & TCSC
1	2211.64	1987.41	1922.78	1908.07
2	2211.64	1987.41	1924.65	1909.94
3	2211.04	1985.67	1922.91	1908.2
4	2211.64	1987.41	1923.91	1909.2
5	2211.64	1987.41	1924.65	1909.94
6	2211.64	1986.22	1923.46	1908.75
7	2205.82	1979.24	1916.48	1901.77
8	2211.64	1987.41	1924.65	1909.94
9	2211.64	1987.41	1918.44	1903.73
10	2211.64	1987.41	1924.65	1909.94
11	2211.39	1987.16	1924.4	1909.69
12	2211.64	1985.91	1923.15	1908.44
13	2211.64	1987.41	1924.65	1909.94
14	2207.87	1983.64	1920.88	1906.17
15	2211.64	1987.41	1924.65	1909.94
16	2211.64	1987.41	1924.65	1909.94
17	2211.64	1987.41	1924.65	1909.94
18	2211.64	1986.22	1923.46	1908.75
19	2211.64	1986.22	1923.46	1908.75
20	2211.64	1986.22	1923.46	1908.75
21	2210.32	1986.22	1910.11	1895.4
22	2211.64	1986.22	1923.46	1908.75
23	2211.64	1986.22	1912.33	1897.62
24	2211.64	1986.22	1923.46	1908.75

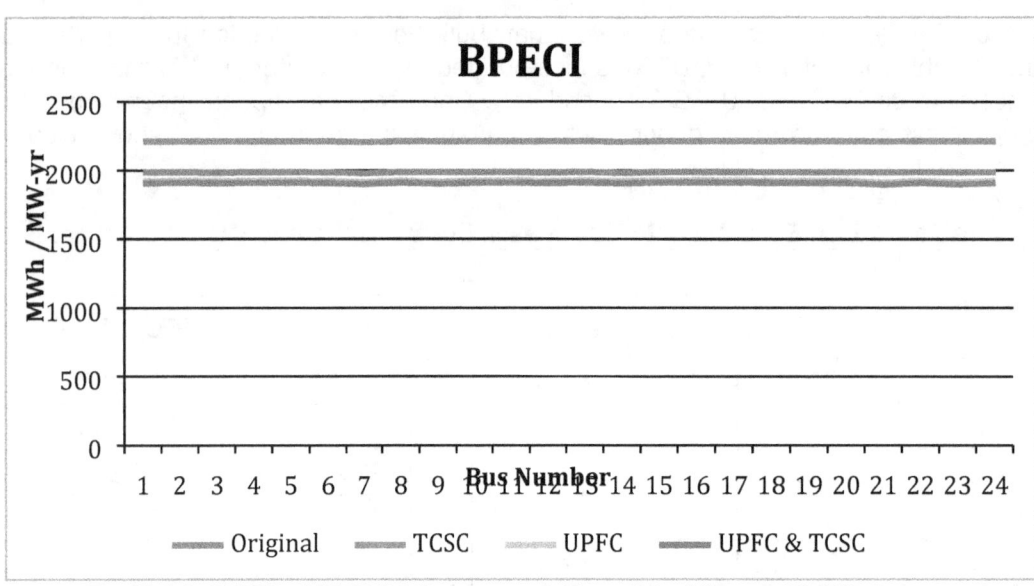

Fig. 7.14: BPECI of IEEE 24 Bus at each Bus with different FACTS Components

From Table 7.11 it can be observed that, Bulk Power Energy Curtailment Index is decreasing when the combination of TCSC & UPFC is incorporated into the system rather than the system when incorporated by TCSC, UPFC independently. Once the Curtailment Index decreases obviously the system performance increases. The graphical form in Fig. 7.14 shows clearly the reduction in BPECI.

7.6.3 Probability of Failure & EENS

Further, Probability of failure & EENS of the system are also calculated at each bus which is presented in Table 7.12 & 7.13 and graphically in Figs. 7.15 & 7.16 respectively.

Table 7.12: Probability of Failure for IEEE 24 Bus RTS at different Buses

Bus No	Probability of Failure			
	Original	TCSC	UPFC	UPFC & TCSC
1	0.0752745	0.0751432	0.0749987	0.0741948
2	0.0752745	0.0751432	0.0749987	0.0741948
3	0.0752745	0.0751338	0.0749587	0.0741348
4	0.0752746	0.0751432	0.0747987	0.0741948
5	0.0752746	0.0751534	0.0749987	0.0741958
6	0.0752749	0.0751432	0.0749987	0.0741548
7	0.0752211	0.0750012	0.0747641	0.0740012
8	0.0752745	0.0751432	0.0749957	0.0741948
9	0.0752745	0.0751232	0.0747987	0.0741978
10	0.0752745	0.0751402	0.0749987	0.0741248
11	0.0752745	0.0751432	0.0749977	0.0741948
12	0.0752745	0.0751132	0.0747985	0.0741948
13	0.0752745	0.0751432	0.0749987	0.0741941
14	0.0752746	0.0751302	0.0748947	0.0741949
15	0.0752745	0.0751432	0.0749981	0.0741748
16	0.0752745	0.0751487	0.0749987	0.0741948
17	0.0752745	0.0751432	0.0748967	0.0741942
18	0.0752745	0.0751432	0.0749987	0.0741448
19	0.0752745	0.0751439	0.0749981	0.0741944
20	0.0752745	0.0751432	0.0749984	0.0741947

21	0.0752745	0.0751431	0.0748987	0.0741748
22	0.0752746	0.0751402	0.0749967	0.0741949
23	0.0752746	0.0751412	0.0749787	0.0740949
24	0.0752745	0.0751432	0.0749987	0.0741948

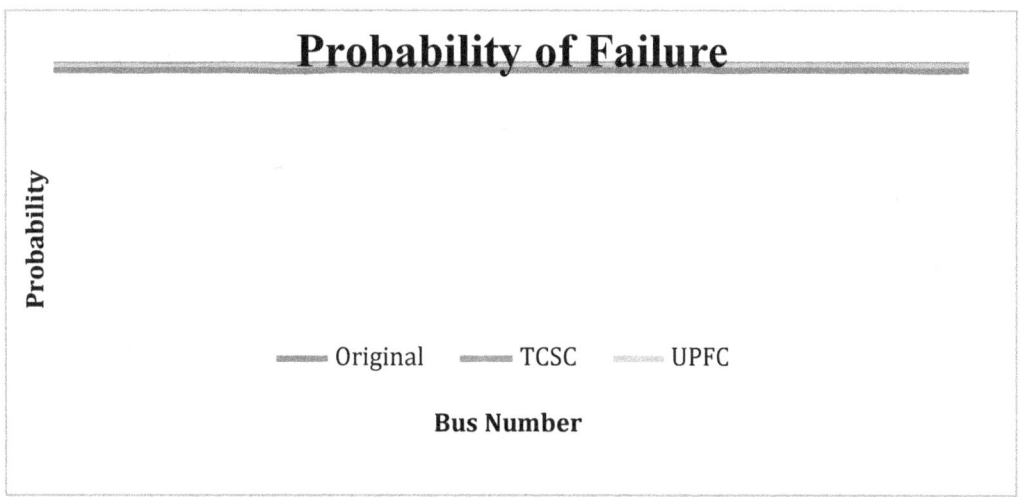

Fig. 7.15: Probability of Failure for 24 bus RTS at different bus

From Table 7.12, it can be observed that, Probability of failure is decreasing when the combination of TCSC & UPFC is incorporated into the system rather than the system when incorporated by TCSC, UPFC independently. Decrease in Probability of Failure indicates increase in the availability of the system, which leads to increase in system performance. The graphical form in Fig. 7.15 shows clearly the decrement of Probability of Failure at each and every bus.

Table 7.13: EENS for 24 Bus RTS at different Buses

Bus No	EENS			
	Original	TCSC	UPFC	UPFC & TCSC
1	3981.03	3802.9	3583.68	3382.41
2	3575.56	3387.43	3168.21	2966.94
3	6635.01	6456.88	6237.66	6036.39
4	2727.83	2549.7	2330.48	2129.21
5	2617.23	2439.1	2219.88	2018.61
6	5013.68	4835.55	4616.33	4415.06
7	4605.1	4426.97	4207.75	4006.48
8	6303.26	6125.13	5905.91	5704.64
9	6450.7	6272.57	6053.35	5852.08
10	7187.92	7009.79	6790.57	6589.3
11	6781.65	6603.52	6384.3	6183.03
12	3198.47	3020.34	2801.12	2599.85
13	9768.18	9590.05	9370.83	9169.56
14	7151.29	6973.16	6753.94	6552.67
15	4684.9	4506.77	4287.55	4086.28
16	3686.14	3508.01	3288.79	3087.52
17	4368.59	4190.46	3971.24	3769.97
18	4274.7	4096.57	3877.35	3676.08

19	6671.88	6493.75	6274.53	6073.26
20	4718.24	4540.11	4320.89	4119.62
21	5719.24	5541.11	5321.89	5120.62
22	3687.19	3509.06	3289.84	3088.57
23	6781.92	6603.79	6384.57	6183.3
24	7014.67	6836.54	6617.32	6416.05

From Table 7.13, it can be observed that, Expected Energy not supplied is decreasing when the combination of TCSC & UPFC is incorporated into the system rather than the system when incorporated by TCSC, UPFC independently. Decrease in EENS indicates increase in the availability of the system, which leads to increase in system performance. The graphical form in Fig. 7.16 shows clearly the decrement of EENS at each and every bus.

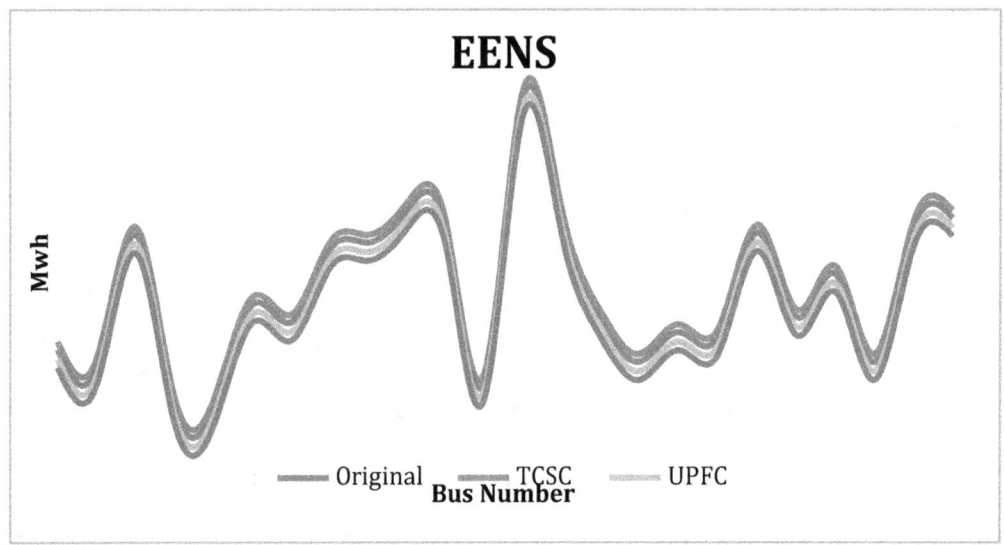

Fig. 7.16: EENS for 24 bus RTS at different bus

7.7 Conclusion

In this chapter, the reliability analysis of sample power systems when using the combination of TCSC & UPFC is presented. Depending upon the generation & transmission line capacity, the combination of TCSC & UPFC is divided into 2 stages. Stage 1, consist 3 Modules each of TCSC & UPFC, where as Stage 2, consists 3 modules of TCSC & 4 Modules of UPFC for 6 bus RBTS. Similarly for 24 bus RTS stage 2 consists 1 module UPFC & 1 Module TCSC, where as stage 1 is same of RBTS. Reliability analysis of the two stages is determined by using state space and series parallel representation. System Indices, Probability of Failure & EENS are also calculated for both the stages.

A comparison has been made between the two stages in all aspects and found that stage 2 is having less supply disturbance, Severity Index, probability of failure, EENS and interruption index when compared with stage 1 in 6 bus RBTS. The availability of stage 1 is greater than stage 2 which is a negligible variation and can be neglected. Finally, it can be concluded that stage 2 can be implemented in to the system depending upon the total power generated, power transfer capacity etc.

In IEEE 24 bus RTS system stage 1 & 2 are incorporated simultaneously depending on the transmission line capacity connected between different buses. System Indices, Probability of Failure & EENS are calculated for all the combinations of FACTS controllers of the system and found the combination of TCSC & UPFC is best suitable for the system rather than other combinations.

A. 6 Bus Roy Billinton Test System Data

Table A1.1: Branch data of 6 Bus RBTS

No.	R pu	X pu	B pu	Failure Rate (f/yr)	MTTR (h)
1,6	0.0342	0.18	0.0212	1.5	10
2,7	0.114	0.6	0.0704	5	10
3	0.0456	0.24	0.0282	2	10
4	0.0228	0.12	0.0141	1	10
5	0.0228	0.12	0.0141	1	10
8	0.0228	0.12	0.0141	1	10
9	0.0023	0.08	0	0.1	768

Table A1.2: Generation Data of 6 Bus RBTS

Bus No.	No. of Units	Rating MW	MTTF (h)	MTTR (h)
1	2	40	1460	45
1	1	10	1752	45
1	1	20	4380	45
2	4	20	3650	55
2	2	5	4380	45
2	1	40	2920	60

B. Data of Transmission System

Table A1.3: Transmission data of the System under study

Transmission	Capacity [MVA]	Line Impedance p.u/km	Failure rate [f/yr]	Repair Time [hrs]
L_1	3000	0.0012+j0.016	1.5	200
L_2	1500	0.0024+j0.019	0.7	150

C. IEEE 24 Bus Reliability Test System Data

Table A1.4: Generation data of IEEE 24 bus System

Type of Generation	Generation Capacity of Each Unit (MW)	No. of Units	Total Generation Capacity (MW)
Oil	12	5	60
Oil	20	4	80
Hydro	50	6	300
Coal	76	4	304
Oil	100	3	300
Coal	155	4	620
Oil	197	3	591
Coal	350	1	350
Nuclear	400	2	800
Total System Generation Capacity:			3405

Table A1.5: Line data of IEEE 24 Bus System

Line No.	Bus I	Bus J	Capacity (MVA)	Forced Outage Probability
1	1	2	193	0.0438
2	1	3	208	0.0582
3	1	5	208	0.0377
4	2	4	208	0.0445
5	2	6	208	0.0548
6	3	9	208	0.0434
7	3	21	510	0.1750
8	4	9	208	0.0411
9	5	10	208	0.0388
10	6	10	193	0.1317
11	7	8	208	0.0342
12	8	9	208	0.0502
13	8	10	208	0.0502
14	9	11	510	0.1750
15	9	12	510	0.1750
16	10	11	510	0.1750
17	10	12	510	0.1750
18	11	13	600	0.0502
19	11	14	600	0.0490
20	12	13	600	0.0502
21	12	23	600	0.0653
22	13	23	600	0.0615
23	14	16	600	0.0477
24	15	16	600	0.0414
25	15	21	1200	0.0515
26	15	24	600	0.0515
27	16	17	600	0.0439
28	16	19	600	0.0427
29	17	18	600	0.0402
30	17	22	600	0.0678
31	18	21	1200	0.0439
32	19	20	1200	0.0477
33	20	23	1200	0.0427
34	21	22	600	0.0565

References

[1] Stephen A. Mallard, Virginia C. Thomas, *"A Method for Calculating Transmission System Reliability"*, IEEE Transactions on Power Apparatus and Systems, Vol. PAS-87, No. 3, March 1968, pp: 824-834.

[2] Roy Billinton, *"Bibliography on the Application of Probability Methods in Power System Reliability Evaluation"*, IEEE Winter Power Meeting, New York, N.Y., 31st Jan. – 5th Feb. 1971, pp: 649-660.

[3] Murthy P. Bhavaraju, Roy Billinton, *"Transmission System Reliability Methods"*, IEEE Winter Power Meeting, New York, N.Y., 31st Jan. – 5th Feb. 1971, pp: 628-637.

[4] A. D. Patton, *"A probability method for bulk power system security assessment, II-Development of probability models for normally operating components"*, IEEE Winter Meeting, New York, Feb. 1972, pp: 2480-2485.

[5] R. N. Allan, *"Bibliography on the application of Probability methods in Power System Reliability Evaluation 1971-1977"*, IEEE Transactions on Power Apparatus and Systems, Vol. PAS-97, No. 6, Nov./Dec. 1978, pp: 2235-2242.

[6] George E Marks, James W McCourt, Robert J Ringlee, Charles L Rudasill, *"Panel Discussion: Designing for Transmission System Reliability"*, 7th IEEE/PES Transmission & Distribution Conference and Exposition, April 1979, pp: 2-3.

[7] J. Endrenyi, P.F. Albrecht, R. Billinton, G.E. Marks, N.D. Reppen, L. Savaderi, *"Bulk Power System Reliability Assessment – Why and How? Part II: How?"*, IEEE Transactions on Power Apparatus and Systems, Vol. PAS-101, No. 9, Sept. 1982, pp: 3446-3456.

[8] J. Endrenyi, P.F. Albrecht, R. Billinton, G.E. Marks, N.D. Reppen, L. Savaderi, *"Bulk Power System Reliability Assessment – Why and How? Part 1: Why?"* IEEE Transactions on Power Apparatus and Systems, Vol. PAS-101, No. 9, Sept. 1982, pp: 3439-3445.

[9] R. N. Allan, R. Billinton, S.H. Lee, *"Bibliography on the application of Probability methods in Power System Reliability Evaluation 1977-1982"*, IEEE Transactions on Power Apparatus and Systems, Vol. PAS-103, No. 2, Feb. 1984, pp: 275-282.

[10] A. P. Meliopoulos, A. G. Bakirtzis, R. Kovacs, *"Power System Reliability Evaluation using Stochastic Load Flows"*, IEEE Transactions on Power Apparatus and Systems, Vol. PAS-103, No. 5, May-1984, pp: 1084-1091.

[11] Luigi Salvaderi, Roy Billinton, *"A comparison between two fundamentally different approaches to composite system reliability evaluation"*, IEEE Transactions on Power Apparatus and Systems, Vol. PAS-104, No. 12, Dec. 1985, pp: 3486-3492.

[12] R. Billinton, *"Criteria used by Canadian Utilities in the Planning and Operation of Generating Capacity"*, IEEE Transactions on Power Systems, Vol. 3, No. 4, Nov. 1988, pp 1488-1493.

[13] M. Th. Schilling, R. Billinton, A. M. Leite da Silva, M. A. El-Kady, *"Bibliography on Composite System Reliability 1964-1988"*, IEEE Transactions on Power Systems, Vol. 4, No. 3, Aug. 1989, pp: 1122-1132.

[14] R.N. Allan, R. Billinton, R.B. Adler, C.C. Fong, G.A. Hinschberger, R.P. Ludorf, M.F. McCoy, T.C. Mielnik, P.M. O'Neill, N.D. Reppen, R.J. Ringlee, L. Salvaderi, *"Bulk System Reliability – Predictive Indices"*, IEEE Task Force on Predictive Indices, APM Subcommittee, IEEE Transactions on Power Systems, Vol. 5, No. 4, Nov. 1990, pp: 1204-1213.

[15] Li Wenyuan, R. Billinton, *"Effect of Bus load uncertainty and correlation in composite system adequacy evaluation"*, IEEE Transactions on Power Systems, Vol. 6, No. 4, Nov. 1991, pp: 1522-1529.

[16] Mario V. F. Pereira, Neal J. Balu, *"Composite Generation / Transmission Reliability Evaluation"*, Proceedings of the IEEE, Vol. 80, No. 4, April-1992, pp: 470-491.

[17] Jianhua Bian, Parviz Rastgoufard, Jack Davey, *"Transmission System Reliability Analysis in Power Systems Reliability Instruction"*, IEEE Proceedings, SSST/CSA 92, 24th South Eastern Symposium and the 3rd Annual Symposium on Communications, Signal Processing Expert Systems and ASIC VLSI Design, 1992, pp: 213-216.

[18] A.D. Patton, S.K. Sing, "*A Transmission Network Model for Multi-Area Reliability Studies*", IEEE Transactions on Power Systems, Vol. 8, No. 2, May 1993, pp: 459-465.

[19] J. Urbanek, R.J. Piwko, E.V. Larsen, B.L. Damsky, B.C. Furumasu, W. Mittlestadt, J.D. Eden, "*Thyristor Controlled Series Compensation Prototype Installation at the SLATT 500KV Substation*", IEEE Transactions on Power Delivery, Vol. 8, No. 3, July-1993, pp:1460-1469.

[20] A.C.G. Melo, M.V.F. Pereira, A.M. Leite da Silva, "*A Conditional Probability Approach to the Calculation of Frequency and Duration Indices in Composite Reliability Evaluation*", IEEE Transactions on Power Systems, Vol. 8, No. 3, August 1993, pp: 1118-1125.

[21] Tarek A.M. Sharaf, Gunnar J. Berg, "*Loadability in Composite Generation/Transmission Power Systems Reliability Evaluation*", IEEE Transactions on Reliability, Vol. 42, No. 3, Sept. 1993, pp: 393-400.

[22] R. J. Ringlee, Chmn, Paul Albrecht, R. N. Allan, M. P. Bhavaraju, R. Billinton, R. Ludorf, B. K. LeReverend, E. Neudorf, M. G. Lauby, P.R.S. Kuruganty, M. F. McCoy, T. C. Mielnik, N.S. Rau, B. Silvmtein. C. Singh, J. A. Stratton, "*Bulk Power System Reliability Criteria and Indices, Trends and Future Needs*", IEEE Transactions on Power Systems, Vol. 9, No. 1, Feb. 1994, pp: 181-190.

[23] R. N. Allan, R. Billinton, A. M. Breipohl, C. H. Grigg, "*Bibliography on the application of Probability methods in Power System Reliability Evaluation 1987-1991*", IEEE Transactions on Power Systems, Vol. 9, No. 1, Feb. 1994, pp: 41-49.

[24] Neso A Mijuskovis, "*Reliability Indices for Electric Power Wheeling*", IEEE Transactions on Reliability, Vol. 43, No. 2, June 1994, pp: 207-209.

[25] E. Weber, B. Adler, R. Allan, S. Agarwal, M. P. Bhavaraju, Roy Billinton, M. Blanchard, R. D'Aquanni, R. Ellis, J. Endrenyi, D. Garrison, C. Grigg, M. Luehmann, J. Odom, G. Preston, N. Rau, N. Reppen, L. Salvaderi, M. Schilling, A. Schneider, A. Vojdani, T. White, "*Reporting Bulk Power System Delivery Point Reliability*", IEEE Transactions on Power Systems, Vol. 11, No. 3, Aug. 1996, pp: 1262-1268.

[26] J. G. Dalton, D. L. Garrison, C. M. Fallon, "*Value Based Reliability Transmission Planning*", IEEE Transactions on Power System, Vol. 11, No. 3, Aug. 1996, pp: 1400-1408.

[27] Roy Billinton, Satish Jonnavithula, "*A Test System for Teaching Overall Power System Reliability Assessment*", IEEE Transactions on Power Systems, Vol. 11, No. 4, Nov. 1996, pp: 1670-1676.

[28] Satish Jonnavithula, Roy Billinton, "*Topological Analysis in Bulk Power System Reliability Evaluation*", IEEE Transactions on Power Systems, Vol. 12, No. 1, Feb. 1997, pp: 456-463.

[29] Roy Billinton, Steve Adzanu, "*Composite Generation and Transmission system Adequacy assessment with time varying loads using a Contingency Enumeration approach*", Conference on Communications, Power and Computing WESCANEX'97 Proceedings, Winnipeg, MB, May 22-23, 1997, pp: 41-46.

[30] A. Jonnavithula, R. Billinton, "*Features that influence Composite Power System Reliability worth Assessment*", IEEE Transactions on Power Systems, Vol. 12, No. 4, November-1997, pp: 1536-1541.

[31] J.M.S. Pinherio, C.R.R. Dornellas, M.Th. Schilling, A.C.G. Melo, J.C.O. Mello, "*Probing the New IEEE Reliability Test System (RTS-96): HL-II Assessment*", IEEE Transactions on Power Systems, Vol. 13, No. 1, Feb. 1998, pp: 171-176.

[32] A. G. Bruce, "*Reliability analysis of Electric utility SCADA systems*", IEEE Transactions on Power Systems, Vol. 13, No. 3, Aug. 1998, pp: 844-849.

[33] R. Billinton, M. Fotuhi Firuzabad, S.O. Faried, "*Power System Reliability Enhancement using a Thyristor Controlled Series Capacitor*", IEEE Transactions on Power Systems, Vol. 14, No. 1, Feb. 1999, pp: 369-374.

[34] R. N. Allan, R. Billinton, A. M. Breipohl, C. H. Grigg, "*Bibliography on the application of Probability methods in Power System Reliability Evaluation 1992-1996*", IEEE Transactions on Power Systems, Vol. 14, No. 1, Feb. 1999, pp: 51-57.

[35] R. Grunbaum, Jacques Pernot, "*Thyristor-Controlled Series Compensation: A State of the Art Approach for Optimization of Transmission Over Power Links*", ABB Review, 5/1999.

[36] M. Pandey, R. Billinton, *"Electric Service Reliability Worth determination in the Nepal Power System"*, Proceedings of 1999 IEEE Canadian Conference on Electrical & Computer Engineering, Alberta, Canada, May 1999, pp: 1465-1470.

[37] IEEE Reliability Test System Task Force of the APM Subcommittee, *"IEEE Reliability Test System - 1996"*, IEEE Transactions on Power Systems, Vol. 14, No. 3, Aug. 1999, pp: 1010-1020.

[38] B. Bak-Jensen, John Bech, C. G. Bjerregaard, P. R. Jensen, *"Models for Probabilistic Power Transmission System Reliability Calculation"*, IEEE Transactions on Power Systems, Vol. 14, No. 3, Aug. 1999, pp: 1166-1171.

[39] L. Geol, C. Feng, *"Well-Being framework for Composite Generation and Transmission System Reliability Evaluation"*, IEE Proceedings, Generation Transmission Distribution, Vol. 146, No. 5, Sept. 1999, pp: 528-534.

[40] M. Fotuhi-Firuzabad, R. Billinton, S. O. Faried, S. Aboreshaid, *"Power System Reliability using Unified Power Flow Controllers"*, IEEE, 2000, pp: 745-750.

[41] A. A. Chowdhury, Don O. Koval, *"Customer-Responsive Bulk Transmission system reliability performance standards for use in a competitive electricity market"*, IEEE 2000, pp: 2057-2062.

[42] K. Audomvongseree, B. Eua Arporn, *"Composite System Reliability Evaluation using AC Equivalent Network"*, Proceedings of IEEE, 2000, pp: 751-756.

[43] M. Fotuhi Firuzabad, Roy Billinton, Sherif Omar Faried, *"Sub-transmission System Reliability Enhancement using a Thyristor Controlled Series Capacitor"*, IEEE Transactions on Power Delivery, Vol. 15, No. 1, Jan. 2000, pp: 443-449.

[44] Roy Billinton, Mahmud Fotuhi Firuzabad, Sherif Omar Faried, Saleh Aboreshaid, *"Impact of Unified Power Flow Controllers on Power System Reliability"*, IEEE Transactions on Power Systems, Vol. 15, No. 1, Feb. 2000, pp: 410-415.

[45] A. A. Chowdhury, Don O. Koval, *"Development of Transmission System Reliability Performance Benchmarks"*, IEEE Transactions on Industry Applications, Vol. 36, No. 3, May/June 2000, pp: 899-903.

[46] M. Fotuhi-Firuzabad, R. Billinton, S. O. Faried, *"Transmission System Reliability Evaluation Incorporating HVDC Links and FACTS Devices"*, IEEE Power Engineering Society Summer Meeting, 2001, pp: 321-326.

[47] Yan Ou, Chanan Singh, *"Improvement of Total Transfer Capability using TCSC and SVC"*, IEEE Summer Meeting, Vol. 1, 2001, pp: 944-948.

[48] John E. Propst, Daniel R. Doan, *"Improvements in Modeling and Evaluation of Electrical Power System Reliability"*, IEEE Transactions on Industry Applications, Vol. 37, No. 5, Sept./Oct. 2001, pp: 1413-1422.

[49] Roy Billinton, Mahmud Fotuhi Firuzabad, Lina Bertling, *"Bibliography on the Application of Probability methods in Power System Reliability Evaluation 1996-1999"*, IEEE Transactions on Power Systems, Vol. 16, No. 4, Nov. 2001, pp: 595-602.

[50] Roy Billinton, Yu Cui, *"Reliability Evaluation of Composite Electric Power Systems Incorporating FACTS"*, Proceeding of the IEEE Canadian Conference on Electrical and Computer Engineering, 2002, pp: 1-6.

[51] Garng. M. Huang, Yishan Li, *"Impact of Thyristor Controlled Series Capacitor on Bulk Power System Reliability"*, Power Engineering Society Winter Meeting, Vol.1, 2002, pp: 975-980.

[52] J. Chen, J. S. Thorp, *"A Reliability study of transmission system protection via a hidden failure DC load flow model"*, Power System Management and Control, Conference Publication No. 488, IEE, 17th – 19th April 2002, pp: 384-389.

[53] Wenyuan Li, *"Incorporating Aging Failures in Power System Reliability Evaluation"*, IEEE Transactions on Power Systems, Vol. 17, No. 3, Aug. 2002, pp: 918-923.

[54] R. C. Bansal, T. S. Bhatti, D. P. Kothari, *"Discussion of Bibliography on the Application of Probability Methods in Power System Reliability Evaluation"*, IEEE Transactions on Power Systems, Vol. 17, No. 3, Aug. 2002, pp: 924.

[55] Armando M Leite da Silva, Luiz A da Fonseca Manso, George J Anders, *"Composite reliability evaluation for large scale Power Systems"*, IEEE Bologna PowerTech Conference, June 23rd – 26th 2003, Bologna, Italy.

[56] Ching-Tzong Su, Chi-Min Lin, Yung-Fu Wang, *"Economic Dispatch and Spinning Reserve scheduling for Generation Transmission Systems"*, IEEE MELECON, 12th – 15th May 2004, pp: 865-868.

[57] Wendai Wang, James M. Loman, Rodert G. Arno, Pantelis Vassiliou, Edward R. Furlong, Doug Ogden, *"Reliability Block Diagram Simulation Techniques applied to the IEEE std. 493 Standard Network"*, IEEE Transactions on Industry Applications, Vol. 40, No. 3, May/June-2004, pp: 887-895.

[58] Roy Billinton, Yifeng Li, *"Incorporating Multi State Unit Models in Composite System Adequacy Assessment"*, 8th International Conference on Probabilistic Methods applied to Power Systems, Iowa State University, Ames, Iowa, Sept. 2004, pp: 70-75.

[59] T. Tran, J. Choi, D. Jeon, J. Choo, R. Billinton, *"Sensitivity Analysis of Probabilistic Reliability Evaluation of IEEE MRTS using TRELSS"*, 8th International Conference on Probabilistic Methods applied to Power Systems, Iowa State University, Ames, Iowa, Sept. 2004, pp: 76-81.

[60] Armando M. Leite da Silva, Leonidas Chaves de Resende, Luiz Antonio da Fonseca Manso, Roy Billinton, *"Well-Being Analysis for Composite Generation and Transmission Systems"*, IEEE Transactions on Power Systems, Vol. 19, No. 4, Nov. 2004, pp: 1763-1770.

[61] Ajit Kumar Verma, A. Srividya, Bimal C. Deka, *"Impact of a FACTS controller on reliability of composite power generation and transmission system"*, Elsevier, Electric Power Systems Research, Vol. 72, Issue 2, Dec. 2004, pp: 125-130.

[62] Sreten Skuletic, Adis Balota, *"Reliability assessment of composite power systems"*, IEEE CCECE/CCGEI, Saskatoon, May 2005, pp: 1718-1721.

[63] Arunachalam M, Ghamandi Lal, Rajiv C G, *"Performance Verification of TCSC control & protection equipment using RTDS"*, 15th PSCC, Liege, session 32, paper 6, 22nd-26th Aug. 2005, pp:1-7.

[64] Jaeseok Choi, Trungtinh Tran, A. (Rahim) A. El-Keib, Robert Thomas, HyungSeon Oh, Roy Billinton, *"A Method for Transmission System Expansion Planning Considering Probabilistic Reliability Criteria"*, IEEE Transactions on Power Systems, Vol. 20, No. 3, Aug. 2005, pp: 1606-1615.

[65] M. Celo, R. Bualoti, *"Integrated Indices that reflects Reliability assessment for Generation and Transmission Network"*, IEEE MELECON 2006, May 16-19, Benalmadena, Spain, 2006, pp: 978-981.

[66] Ming Zhou, Gengyin Li, Peng Zhang, *"Impact of Demand Side Management on composite Generation & Transmission System Reliability"*, IEEE PSCE 2006, pp: 819-824.

[67] T.X. Zhu, *"A New Methodology of Analytical Formula deduction and Sensitivity analysis of FENS in Bulk Power System Reliability Assessment"*, IEEE PSCE 2006, pp: 825-831.

[68] A. A. Chowdhury, D. O. Koval, *"Probabilistic Assessment of Transmission System Reliability Performance"*, IEEE Power Engineering Society General Meeting, 2006, pp: 1-7.

[69] Fang Yang, A. P. Sakis Meliopoulos, George J. Cokkinides, Q. Binh Dam, *"Bulk Power System Reliability assessment considering protection system hidden failures"*, iREP symposium-Bulk Power System Dynamics & Control, 19th – 24th Aug. 2007, Charleston, SC, USA.

[70] Klaus Habur, Donal O'Leary, *"FACTS – For Cost Effective and Reliable Transmission of Electrical Energy"*, 2008, pp: 1-11.

[71] Sayed-Mahdi Moghadasi, Ahad Kazemi, Mahmud Fotuhi-Firuzabad, Abdel-Aty Edris, *"Composite system reliability assessment incorporating an Interline Power Flow Controller"*, IEEE Transactions on Power Delivery, Vol. 23, No. 2, April 2008, pp: 1191-1199.

[72] Rosario Toscano, Patrick Lyonnet, *"On-Line Reliability Prediction via Dynamic Failure Rate Model"*, IEEE Transactions on Reliability, Vol. 57, No. 3, Sept. 2008, pp: 452-457.

[73] Hamid R. Bay, Ahad. Kazemi, *"Reliability evaluation of composite electric power systems incorporating STATCOM & UPFC"*, IEEE Power & Energy Engineering Conference, APPEEC 2009, Asia-Pacific, 27th – 31st March 2009, pp: 1-6.

[74] Roy Billinton, Robert J. Ringlee and Allen J. Wood, *Power System Reliability Calculations*, The MIT Press, Cambridge 1st Edition, 1973.
[75] Narain G. Hingorani & Laszlo Gyugyi, *Understanding FACTS, Concepts and Technology of Flexible AC Transmission Systems*, IEEE Press, Standard Publications, 1st Indian Edition, 2001.
[76] J. Endrenyi, *Reliability Modeling in Electric Power Systems*, John Wiley & Sons, A Wiley – Interscience Publications, 2002.
[77] Roy Billinton, Ronald N. Allan, *Reliability Evaluation of Engineering Systems*, Plenum Press, New York, 1994, Reprinted in India, BS Publications, 2007.
[78] Roy Billinton, Ronald N. Allan, *Reliability Evaluation of Power Systems*, 2nd Edition, Plenum Press, New York, 1996, Reprinted in India, BS Publications, 2007.

www.ingramcontent.com/pod-product-compliance
Lightning Source LLC
Chambersburg PA
CBHW080931170526
45158CB00008B/2241